SEA FLOOR DEVELOPMENT: MOVING INTO DEEP WATER

SEA FLOOR DEVELOPMENT: MOVING INTO DEEP WATER

A ROYAL SOCIETY DISCUSSION
ORGANIZED BY
SIR ANGUS PATON, F.R.S., SIR PETER KENT, F.R.S.,
SIR GEORGE DEACON, F.R.S.,
SIR KENNETH HUTCHISON, F.R.S.
AND M. B. F. RANKEN
IN COLLABORATION WITH
THE BRITISH NATIONAL COMMITTEE ON
OCEAN ENGINEERING OF THE COUNCIL OF
ENGINEERING INSTITUTIONS

HELD ON 1 AND 2 JUNE 1977

LONDON
THE ROYAL SOCIETY
1978

Printed in Great Britain for the Royal Society
at the
University Press, Cambridge

ISBN 0 85403 100 6

First published in *Philosophical Transactions of the Royal Society of London*,
series A, volume 290 (no. 1366), pages 1–189

Published by the Royal Society
6 Carlton House Terrace, London SW1Y 5AG

PREFACE

All great engineering ventures need scientific data on which to base decisions. Too often such decisions have to be made in situations where the data are lacking or incomplete; this causes uncertainties and delays.

The Council of the Royal Society submitted in April 1974 a memorandum on seabed engineering to the subcommittee of the Select Committee on Science and Technology dealing with that subject. This memorandum stressed the need for continued and increased support for basic studies of the natural forces of the seabed to back up the technical advances being made in deep-sea exploration and exploitation, and the value of close cooperation between scientists and technologists.

The subject was considered by the Industrial Activities Committee, which decided that a discussion meeting covering the problems connected with sea floor development would be appropriate as it would provide an opportunity to bring together the technologists and scientists and offer an opportunity to demonstrate the interactions between engineering and science in the exploitation of the seabed.

The British National Committee on Ocean Engineering of the Council of Engineering Institutions was known to have considerable interest in this subject and consultations were held with this Committee and other interested bodies.

As it was believed that most of the important problems in engineering had been solved for depths down to about 300 m it was decided that the most useful subject for a meeting would be 'Sea floor development: moving into deep water', the term 'deep water' being used broadly to cover problems connected with a depth range of 300–2000 m.

This gave an opportunity of demonstrating British initiative and competence in deep water activities, in particular covering the morphology and currents of the continental margins, subsea engineering, hydrocarbon potential, the geochemistry of ferromanganese deposits, environmental aspects, as well as the up-to-date position on the law of the sea and the ownership of the ocean floor.

The papers submitted at the two-day meeting held on 1 and 2 June 1977 are reproduced in this volume together with a record of the discussion.

March 1978   Angus Paton

CONTENTS

[Seven plates]

PAGE

PREFACE   V

J. BIRKS
Introduction to aspects of economics and logistics   3
Discussion: G. L. HARGREAVES, E. G. WEST   17

SIR ROGER JACKLING
The law of the sea and the deep seabed   21

H. R. WARMAN
Hydrocarbon potential of deep water   33

S. E. CALVERT
Geochemistry of oceanic ferromanganese deposits   43
Discussion: D. S. CRONAN   72

A. S. LAUGHTON AND D. G. ROBERTS
Morphology of the continental margin   75

W. J. GOULD
Currents on continental margins and beyond   87

E. C. GOLDMAN
Offshore subsea engineering   99
Discussion: T. S. McROBERTS, J. BLACK   110

W. H. VAN EEK
The challenge of producing oil and gas in deep water   113

K. B. SMALE-ADAMS AND G. O. JACKSON
Manganese nodule mining   125
Discussion: B. WHITE   133

R. F. BUSBY
Engineering aspects of manned and remotely controlled vehicles   135
Discussion: C. KUO, M. W. THRING   150

R. E. ENGLAND
The underwater contractor: his rôle and development   153

T. H. HUGHES
Effect of the environment on processing and handling materials at sea   161
Discussion: B. WHITE   176

T. F. GASKELL
Environmental pollution in offshore operations   179
Discussion: B. WHITE   185

INDEX   187

A NOTE ON UNITS

A number of non-SI units, still widely used in this subject and therefore in this volume, are listed below with their SI equivalents.

$$\text{inch (in)} = 2.54 \text{ cm}$$
$$\text{foot (ft)} = 0.3048 \text{ m}$$
$$\text{fathom} \approx 1.829 \text{ m}$$
$$\text{nautical mile} \approx 1.853 \text{ km}$$
$$\text{knot} \approx 0.515 \text{ m/s}$$
$$\text{lbf/in}^2 \text{ (p.s.i.)} \approx 6895 \text{ Pa (or 6895 N/m}^2)$$
$$\text{atmosphere} \approx 101 \text{ kPa}$$
$$\text{barrel} \approx 0.159 \text{ m}^3$$
$$\text{barrel oil equivalent} \approx 6.1 \text{ GJ (or } 6.1 \times 10^9 \text{ J)}$$

Phil. Trans. R. Soc. Lond. A. **290**, 3–19 (1978)   [3]

Printed in Great Britain

Introduction to aspects of economics and logistics

By J. Birks

*The British Petroleum Company Limited, Britannic House, Moor Lane,
London EC2Y 9BU, U.K.*

During the past five years the oil industry has moved its exploration and development programmes into progressively deeper waters, so that production operations in 150 m (500 feet) of water are becoming conventional, and exploration in water depths of over 300 m (1000 feet) commonplace.

The first part of this introductory paper is devoted to areas of opportunity in the deeper waters of the sedimentary basins of the world, with particular emphasis on the technical merits of these areas, and the size and high productivity necessary to justify their development. A description follows of the trend in licensing terms, the tax and financial arrangements that might apply, and the growing involvement of national oil companies and national energy policies with their consequent effect on the control of developments, right to export oil, and the division of profits.

The increasing importance of logistic and environmental factors on the technological requirements both in exploration and development is outlined, and some examples drawn of their political and sociological impacts. The development of supporting infrastructure in remote environments, of national preference for materials and services, codes of practice and further constraints in the overall capital investment programmes, are also outlined.

The final section deals with the economic implications of these international activities where during the course of the next 25 years it is expected that offshore oil production rates will double. The nature of the risk investments where exploration wells now cost between £3 and £5 M each, and capital costs for individual projects are over £1000 M, are examined, reflecting differences between the private sector objectives and national oil company objectives. Examples can be drawn from events in O.P.E.C. areas during the past five years.

Introduction

During the past five years the oil industry has moved its exploration and development programmes into progressively deeper waters. The extent of industry's interest may be gauged from the fact that all major companies have extensive deep-water licences some of which carry drilling obligations, as shown in figure 1. The gaps in the licensing pattern shown are partly due to unpromising geology but frequently due to licensing constraints or unfavourable licence terms.

An effective overall strategy for the exploration for and development of commercial hydrocarbons in deep water depends on a careful appraisal of the oil potential and technological and economic factors.

Geology

In the first place I intend to review in brief the critical geological and geochemical factors that determine the prospectivity of these regions – factors which are covered in more detail by Warman (1978, this symposium). The three principal factors are that

(*a*) the sediments must have generated hydrocarbons in sufficient quantity;

(*b*) there must be reservoir rocks of sufficient thickness, porosity and permeability; and

(*c*) structures must be large enough to justify exploitation.

Insufficient is known at present about the question of hydrocarbon generation in deep water. The governing factors must be essentially similar to those on the shelf. We require an adequate supply of organic carbon within the sediments, and this must be heated to a sufficiently high temperature, primarily through burial, to achieve the generation threshold. It is apparent that the type of kerogen, the insoluble organic matter, plays a large part in determining whether oil or gas is generated, humic kerogen of terrestrial origin being essentially gas prone, and the presence of organic carbon of marine origin being necessary for the generation of oil.

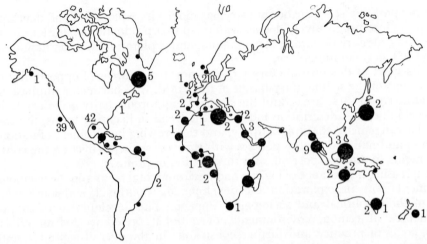

FIGURE 1. Deep-water activity at more than 200 m, April 1977. Before 1973, 36 wells were drilled, in 1973 the number was 5; in 1974, 10; in 1975, 33; in 1976, 58; in 1977 (to April), 17; total, 159. Circle size represents licence areas in thousands of square kilometres: the smallest less than 10, the next between 10 and 50, the next between 50 and 100, and the largest over 100.

Until recently it was therefore believed that there was a greater likelihood of oil rather than gas being generated from the essentially marine sediments found in the deep-water environment. Recent studies, however, have served to emphasize the importance of reducing or oxygen-deficient conditions for oil generation. Such conditions are far removed from the dynamic, well oxygenated, ocean waters that flow along the majority of present oceanic margins, but may be sought within basins of restricted circulation associated with the initial phases of ocean formation.

Consideration of average geothermal gradients and control from boreholes indicates that throughout much of the world a minimum sedimentary thickness of approximately 3 km is required to reach the hydrocarbon generation threshold of approximately 70 °C, although the most recent results from the Deep Sea Drilling Project suggests that this threshold may be attainable at a depth of as little as 1.5 km. Increasingly detailed information from the floors of the oceans confirms that the necessary thick sedimentary accumulations are present almost exclusively along the edges of the major land masses, from which they have been derived. The abyssal depths which comprise nearly 80% of the oceans have smaller thicknesses of sediment and are unlikely to yield commercial hydrocarbons (figure 2). With few exceptions our attention is thus concentrated upon the continental margins.

We must next consider possibilities for the occurrence of suitable reservoir rocks within this marginal sedimentary wedge or prism. It should be stressed that, with the exception of either fracturing or secondary porosity developed in chalks or other very fine grained lithified oozes, all reservoir rocks are ultimately of shallow-water origin, or are derived from a shallow-water source. Even the so-called deep-sea fan sands come initially from an erosional source either on land or in shallow water. Very few holes have been drilled to any depth on the continental slope. Some control in deeper water is available, however, from holes drilled by the Deep Sea Drilling Project. These confirm that the bulk of sediments present in the ocean basins and on

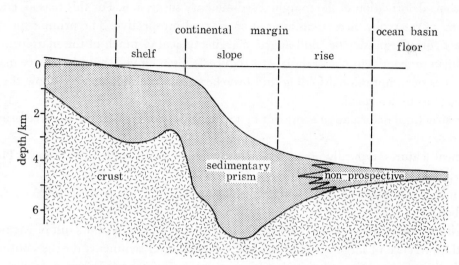

FIGURE 2. Schematic cross section through a passive continental margin.

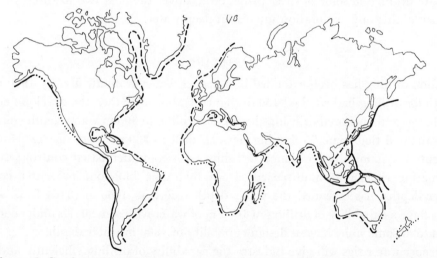

FIGURE 3. Continental margin types: ——, subduction; ·····, transform; ---, pull apart.

the continental rises are fine-grained with low matrix permeability and are thus totally unsuitable as hydrocarbon reservoirs. Moving from the shelf into deeper water, we therefore expect an overall deterioration in reservoirs, both in quality and quantity. Whether this deterioration is gradual or abrupt will depend upon local circumstances. We have much to learn yet about the mechanisms whereby essentially shallow-water sediments, whether siliciclastics or carbonates, may be deposited in deeper water.

The other alternative is to seek reservoirs within the older rocks of continental margins that predate the formation of an ocean basin. These may include sediments of shallow-water origin subsequently depressed into deeper water as a result of continued marginal subsidence and/or sediment loading. Unfortunately, in many cases these prospects have been buried beneath later sedimentary accumulations at too great a depth for there to be much chance of extraction, if not of the existence, of commercial hydrocarbons.

The final factor to be considered in this geological résumé is that of structural deformation. Certain areas in the immediate vicinity of subducting margins typified by those encircling the Pacific show deformation of the marginal sedimentary succession. For this reason, coupled with problems of reservoir, these areas are not considered prospective. The prime exploration areas are therefore the passive or 'pull-apart' margins typical of much of the margins on the Atlantic, Indian subcontinent, Australasia, etc. (figure 3). By their very nature, passive margins are not the sites of compressional folding and over large areas seismic sections show the sedimentary wedge as undeformed.

However, structural deformation is present in a wide variety of areas and is of three principal types related to:

(i) basement features, typically horsts bounded by faults, for example the Exmouth Plateau of Australia;

(ii) salt or shale diapirs, associated with faults, for example in various areas off West Africa, such as Gabon;

(iii) gravity tectonics, including growth faults and simple folds as in the Gulf of Mexico.

Despite the presence of large areas of relatively undisturbed sediments on the continental slope and rise it is encouraging to note that large structures, often simple and domal, do exist and results of drilling on some of these prime targets over the next two to three years will be critical in sustaining any exploration interest in deep water.

DRILLING

Exploration drilling has been extended into deeper water dramatically. By 1970 the record water depth for drilling had reached 456 m (figure 4), but since then the development of semi-submersible rigs and anchored drillships has made drilling to 600 m water depth commonplace and it is estimated that some 75 floating rigs can drill to 300 or more metres. However, the development of dynamically positioned drillships with computer based control systems, with buoyant marine risers and multiplexed electro-hydraulic blow-out preventer controls for great water depths, has extended the water depth record to 1055 m (3461 feet). At present some 8 rigs have a capability of drilling in 1000 m of water with some 6 rigs under construction, and of this total some units have a design capability of 1800 m water depth.

These deeper water rigs will give industry the capability of drilling efficiently and safely for commercial hydrocarbons over the upper continental slope.

PRODUCTION

I should now like to refer to some of the problems related to the practical and environmental aspects of field development in deeper water. Initially it is perhaps appropriate to review the degree of multidisciplinary skills that are involved in one way or another within the overall development requirements (table 1). Fortunate is the company that has such wide and diverse skills entirely within 'in house' resources. The majority of smaller operators will require to use consultants as and when necessary and BP has recognized this potential in forming a Technical Services Group for the purpose of making such expertise available on a commercial basis.

At present, offshore development by means of fixed platforms has been extended to 200 m for the North Sea and 300 m in the quieter water of the Gulf of Mexico. From studies undertaken by BP it is believed that at these depths of water we are at, or close to, the economic limit of such conventional platforms. It is, of course, relevant to add that the economic limit will be influenced by several factors, including the size and shape of the reservoir and also the well productivity. Additionally, the increasing momentum of subsea wellhead development

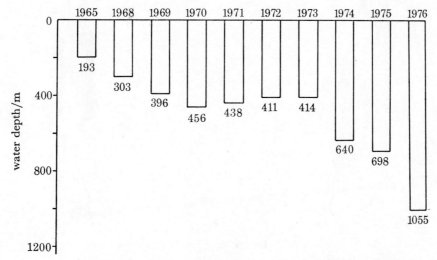

FIGURE 4. Annual deep-water drilling records (industrial operators).

TABLE 1. APPRAISAL FOR DEVELOPMENT PLANNING IS AN ACTIVITY
INVOLVING MANY DISCIPLINES

the sciences	engineering	other
biochemistry	chemical	accountancy
botany	civil	cartography
chemistry	control	computer sciences
fluid mechanics	drilling	economics
geochemistry	electrical	environmental sciences
geology	electronic	ergonomics
geophysics	mechanical	finance
marine biology	metallurgical	law
mathematics	petroleum	management sciences
meteorology	production	medicine
oceanography	telecommunications	psychology
palaeontology		satellite navigation
palynology		social sciences
physics		surveying
statistics		

will tend to extend the application of conventional platforms and provide a springboard for deeper water technology. For the purposes of this review I propose to emphasize the trend and problems related to water depths greater than 200 m and for practical purposes limit the range to 650 m, which covers the new systems now under active development.

For the immediate future I would envisage offshore oilfield systems retaining facilities at the surface to house operational personnel and the major process and power facilities. Within this general concept considerable development work is under way in the field of tethered buoyant

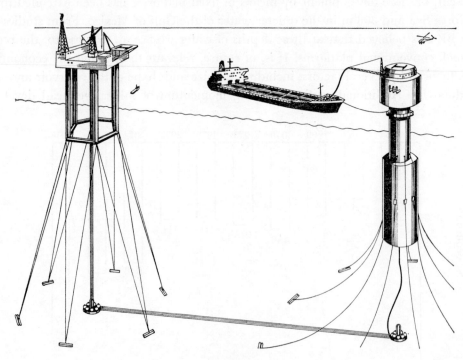

FIGURE 5

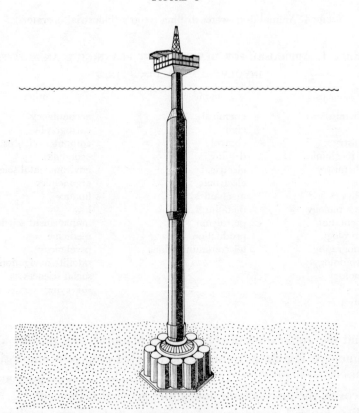

FIGURE 6. Arcolprod (Articulating Buoyant Column Production and Storage System).

platforms (t.b.ps) (figure 5), articulated columns (figure 6) and stayed tower (figure 7). All of these concepts are sensitive to payload and accordingly the development of efficient equipment for processing, power generation and effluent clean-up, requiring less space and of considerably reduced weight, will be of considerable importance to ultimate system economics.

Conventional systems so far installed have relied predominantly on the drilling of wells from a platform after placement at the field. This has necessitated concurrent drilling and production operations and reliable and safe procedures for accomplishing this have been developed. Notwithstanding the excellent record of the industry for undertaking such operations safely and efficiently, the Norwegian Petroleum Department have introduced (mid-1976) new regulations seeking to segregate these two operations which, if pursued, can only tend to have an adverse economic effect on the scope of conventional systems.

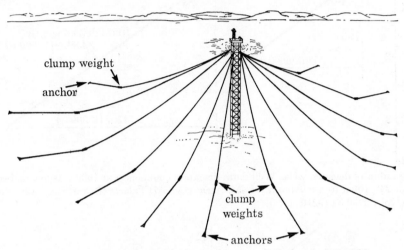

FIGURE 7. Schematic of Exxon stayed tower; deck area is 45 000 ft².

With such developments as tethered buoyant platforms there is opportunity to undertake drilling before installation of the platform, either directly below the platform, in which case the number of wells will by programme restraints be limited to 12–15, or alternatively at a number of well clusters located at strategic locations to allow efficient drainage and optimum deviation related to pay zone depth (figure 8). Well clusters of this form, say six wells per cluster, would have the advantage of limiting the number of producing wells to be closed in during workover (well maintenance) operations but would suffer the disadvantage of wells being remote from the platform for the more frequent service tasks normally undertaken by wireline on land or conventional platform wells. Pump-down tools, more frequently referred to as t.f.l. (through flow line) have been developed to undertake such work but the majority of current development and experience has been restricted to a tubing size of $2\frac{1}{2}$ in as applicable to lower productivity wells. For high producers such as are found in the North Sea, and will be required for viable deeper water developments, well completion requirements will necessitate 4 in, or perhaps larger, t.f.l. tools. Operation of such tools from a manifold chamber located at the well cluster will also require consideration as despite the added problems, the possible savings of potential well losses can be considerable. Figure 9 indicates the potential losses of production through twin 3 in and 4 in flowlines at various well flow rates at distances exceeding 3–4 km from the process facility. In total terms the back pressure on the well production system consists of losses in the well tubing, which is therefore greater for deviated wells and smaller tubing size; the length and size of flowline; and the lift to surface which relates to water depth. As this restraint increases with deeper water, consideration will need to be given to first-stage separation at the sea bed, which will also have the advantage of extending the range of flow without incurring surging, i.e. gas and oil separating out in the pipeline in the form of slugs which can cause considerable problems in the process system.

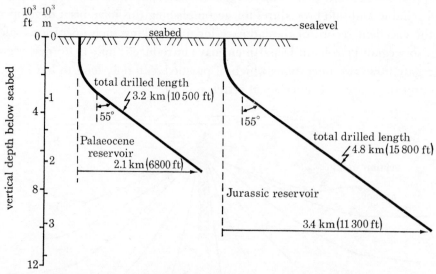

FIGURE 8. Reach radius of deviated wells. Well data: deviation commences at 1000 ft below seabed and the angle is built up at $2\frac{1}{2}°/100$ ft to a maximum of 55°. Areas covered: Palaeocene reservoir, 1360 ha (3360 acres); Jurassic reservoir, 3730 ha (9210 acres).

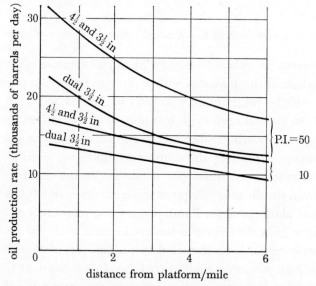

FIGURE 9. Typical North Sea oilfield peak production. $10\frac{3}{4}$ in casing: $4\frac{1}{2}$ and $3\frac{1}{2}$ in tubing and flowlines; $9\frac{5}{8}$ in casing: dual $3\frac{1}{2}$ in tubing and flowlines.

Of the systems currently under development it is reassuring to note that the basic operating techniques at the process facility are not too different from existing technology and hence the majority of experience gained in safety involving personnel and the environment will still apply. The same is not true of riser systems for oil and gas, which flex continuously with the horizontal excursion of t.b.ps, and considerable research and development is now in process towards resolving these problems. Also the range and scope of work that can be performed using subsea intervention systems is still the subject of active review, mainly related to the safety of personnel involved and safety of the environment rather than the basic concepts which have already been established in such systems as developed by SEAL and Lockheed and under active development by Vickers Intertek. As we extend technology into deeper water it is inevitable that we will face greater problems related to the logistics of supply for both material and human resources necessary to operate and maintain the field installation.

Despite the considerable problems already faced and overcome in the North Sea and the Gulf of Mexico developments, their close proximity to fully developed industrial resources has been of major significance. In the case of the North Sea the industrial infrastructure existed before the event and the essential requirement was to build the necessary oil experience to it. Even this more limiting requirement has taken time to establish but it does help to illustrate the depth of problem that will exist when undertaking such development in the off-shore areas of underdeveloped nations.

For such underdeveloped areas of the world, whose offshore area will considerably exceed that of developed areas, the development of a supporting infrastructure will need to feature in the very early development phases following identification of exploitable crude reserves. The oil industry has considerable experience from its past history related to earlier operations in the Middle East, but it is equally true that until fairly recent times the political and sociological impacts were of a much smaller degree than would apply today. Whereas in the past the tendency was for the operating company to develop its own infrastructure of industrial and social services within its field of operations and import the necessary materials and equipment, the requirements of today demand the maximum involvement of national resources together with the active participation of the host government. As the nature of offshore operations is to minimize the amount of work necessary at the offshore installation, e.g. maintenance as far as possible by unit replacement methods, it will be readily appreciated that the development of such an infrastructure may well incur more time and effort than the objective of developing an oilfield. Also to be included in such an infrastructure would be the vital service companies related to the supply of workover facilities, subsea intervention, and inspection services, and such companies would also need to be mindful of the aspirations of the host government, both with regard to the involvement of national investment opportunities and the training of nationals towards staffing such enterprises at all levels. As an example of these developments one need only review the changes that have taken place in such areas as Iran, Abu Dhabi, Qatar and Kuwait where in the course of the last two decades the rôle of the oil company has moved towards that of a service company rather than the earlier equity-based crude producer operating from within its own supporting infrastructure.

As the experience and confidence of national oil companies increases so will the tendency to introduce local codes of practice covering materials and services. Such a tendency could be dangerous and difficult to administrate if introduced in an indiscriminate and unprofessional manner and it is to be hoped that such will not occur. Certainly there is every reason to support codes of practice relevant to the geographical area of application and it is readily apparent that what is suitable for the placid waters of the Persian Gulf is not relevant to the North Sea, but, equally, divergence of regulations by different Governments for the same environmental conditions can only result in a degree of confusion. The objectives of the International Exploration and Production Forum formed in 1974 to represent the industry on an international basis will I hope successfully overcome any major problems in this respect by their close cooperation with governments and National Operating Committees.

ECONOMIC ASPECTS OF DEVELOPMENT

With increasing offshore development activity, and in particular as we move into deeper and more hostile environments, field development proposals must include more rigid economic appraisal in the early planning phases.

It will be readily appreciated that as recently as ten years ago the majority of oilfield development was still on land and the capital cost per barrel per day was of the order of hundreds of pounds whereas current offshore development is in thousands of pounds per barrel per day capacity. Additionally, risk factors related to the successful completion of offshore projects are correspondingly higher with regard to finance, programme time and environmental hazards.

Factors which have a bearing on development economics are shown in table 2.

TABLE 2. FACTORS AFFECTING DEVELOPMENT ECONOMICS

environment	water depth
	sea state
	seabed soil mechanics
reservoir	volume of recoverable reserves
	transmissibility
	shape
	pay depth
	thickness
	configuration
	crude quality
type of oil outlet	pipeline or marine terminal
timing of production	
licence terms	
crude pricing	

Water depth and environmental conditions determine the design requirements of the production platforms, with a move from gravity or piled bottom founded structures to guyed towers or tethered buoyant platforms related to economics. Not only are increased construction costs incurred in deeper water but installation times may be longer and the penalties of missing seasonal weather windows may be catastrophic.

The shape of the reservoir and the drilled depth, which affects well deviation, have a bearing on the reservoir volume that can be drained effectively from one platform: for a given volume of recoverable reserves a long narrow field with a thin reservoir section will require more platforms for effective drainage. Remote subsea well completions, tied into a central platform, will improve the drainage achieved by one production complex.

The volume of recoverable reserves and reservoir transmissibility are of paramount importance in determining the peak sustainable production rate and number of production wells required. Field development economics are sensitive to well numbers, particularly in deep water because of the high costs of deep capability rigs, and subsequent well maintenance.

Reservoir and crude-oil characteristics are important in determining process type requirement.

For land-based development there is no problem in extending the production facility as the knowledge of the reservoir increases, but for offshore development it is vital to include for all eventual requirements within the original project if the most economic development is to result. Within the information available from appraisal drilling the prediction of all reservoir characteristics and future behaviour is no easy task.

Naturally all these factors must be taken into account when planning a field development and extensive sensitivity testing must be applied to determine those factors that have most impact on the overall economics, in order that the programme and development planning can be tailored to minimize the effect of the key variables.

To illustrate this point, economic evaluation often demonstrates very forcibly the advantages of planning development to give production at the earliest possible date following financial commitment. Such a situation is difficult to achieve from platform drilling as there are definite limitations both technically and logistically to the number of rigs that can operate from a platform. Subsea completions will in the near future provide the most effective solution to the early production requirement, albeit that the maintenance of such subsea completions will be technically more difficult and more costly than conventional platform wells.

Having said all this, it is apparent that each field discovery will be subject to its own set of development conditions and it is not practicable to make really significant generalizations which relate water depth to economic reserve size. Nevertheless, one can very broadly say that under many existing tax régimes, and assuming crude prices remain constant in terms of purchasing power, then large recoverable reserves (of at least 5×10^9 barrels, i.e. giant field) allied to high and sustainable individual well flow rates (5000 barrels per day or better) appear necessary if fields in around 500 m water depth will be economically viable.

This economic assessment has feedback to the exploration geology outlined earlier. It is evident that the very factors which are desirable in deep water, namely good transmissibility and large reserve size, will be elusive. Fields of the required size will be restricted in number (giant fields are estimated to represent less than 1 % of the world's fields but account for 75 % of global reserves), and therefore any method of production, such as subsea well completions, and any improvement in licence terms such as those discussed in the next section, will encourage the search for smaller fields and their exploitation.

LICENCE TERMS

I have identified the factors affecting the size of field, the productivity necessary to justify development in deeper waters, the engineering and environmental aspects, but not least among these are licensing, tax and other financial terms. In the face of the growing economic and political significance of oil there has been a general tightening of terms in recent years on a world-wide basis. The results of this are increased government take, direct or indirect control of activities and state participation but only rarely with State involvement in exploration risk.

Terms vary enormously but for the most part fall into three broad categories:

(1) Royalty/tax terms as we have in the U.K. and Norwegian sectors of the North Sea. Under these terms, government take could be some 75 % or more.

(2) Production-sharing terms such as apply in Indonesia and Malaysia, generally with production split after cost recovery, in the range 70:30 to 85:15 in favour of the State.

(3) Service contracts.

State participation without involvement in exploration risk, superimposed on these terms, is a further burden. Participation can be as high as 75 % in certain circumstances.

Any state requirement to defer development of discoveries, restrict production or statutory variations of terms during the life of a project can have severe effects on the economics as far as the companies are concerned and can undermine confidence in continued exploration.

At present, terms rarely differentiate between developments in shallow and deep water although in some cases discretionary provisions exist to reduce royalty and/or tax to encourage development of marginal fields.

However, increasing activity in deeper water with the much higher exploration and development costs and considerably greater risk justifies special consideration. Perhaps what are required under either royalty/tax or production-sharing terms to provide the necessary incentives for exploration and development are:

(1) Bonus bidding should be replaced by committed exploration programmes, so that maximum investment is made in exploration, but there should be more discretion about such commitments and onerous requirements should be avoided.

(2) Depreciation rates and other financial provisions should be such that costs are recovered over as short a period as possible in order that the companies can recycle cash into further exploration and development.

(3) The terms should be flexible to ensure that the economic cut-off for field development is as low as is realistically possible and such flexibility should be statutory and not discretionary.

(4) Reasonable assurance is needed that terms will be stable during the life of a project and not subject to arbitrary change after investment decisions have been taken.

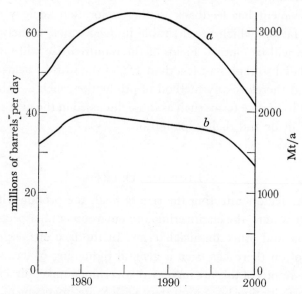

FIGURE 10. Non-Communist world crude oil availability. (*a*) Total crude production; (*b*) O.P.E.C. production range at full production capacity.

THE INTERNATIONAL SCENE

Energy forecasters in the oil industry, government agencies, and academic institutes are unanimous that the world will need additional oil supplies, at least for the next decade. Equally, they predict a large increase in the use of gas as new supplies are developed and moved internationally. However, there is now wide acknowledgement that world oil production is likely to reach a peak some time in the 1980s, the precise timing depending on the policies of the governments of the major oil producing countries – including the United Kingdom and Norway as well as Saudi Arabia and Kuwait – and also of course on new discoveries. However rapidly alternative energy sources such as coal, nuclear, hydro and solar are developed, because of the technical lead time and the practicality of projects, major new supplies from these sources will not be available to meet increased energy demand for at least ten years.

Figure 10 gives an illustration of the crude-oil availability to the non-Communist world and the part that the Organization of Petroleum Exporting Countries (O.P.E.C.) may play, assuming that new oil reserves to be discovered and developed in the period will provide about one-third of the production potential in the year 2000, and O.P.E.C. oil prices as the price setter will be at least maintained in real terms. Peak availability in the second half of the 1980s emphasizes the need for intensive oil exploration programmes.

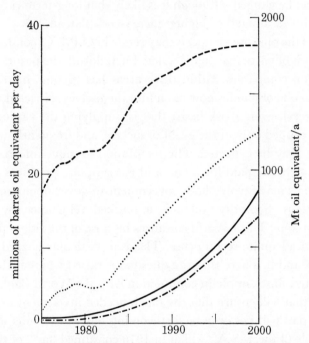

FIGURE 11. Non-Communist world natural gas potential, production and trade: ---, Non-Communist world production potential; ·····, O.P.E.C. production potential; ——, total international gas trade; liquefied natural gas trade.

Figure 11 illustrates an even more dramatic development of gas reserves, where it is assumed that new gas discoveries will provide well over half the production potential in the year 2000. Apart from the indication of a tremendous rate of growth of the international liquefied natural gas trade (the majority from O.P.E.C. areas), in energy terms there may be as much gas required as oil to meet the free world requirements by the year 2000.

For the Communist world, including China, as indicated by a recent Central Intelligence Agency report, and on our appreciation, oil production potential may peak well before 1990 at about 15×10^6 barrels per day, sharing common problems with the free world.

The prospects offshore within this overall scenario, however, are the most promising with production potential from new discoveries forecast to be at least equal to the current potential of about 10^7 barrels per day.

It is not surprising that offshore exploration and production investments have more than doubled in real terms in the past three years, and are expected to increase in the foreseeable future. In BP's own case, 80 % of its exploration investment is offshore in mainly non-O.P.E.C. areas, and its exploration and production investments for oil, gas, coal and minerals account for a half or more of its total investments in the next 5 years.

The rapid increases in unit development costs offshore are alarming, and pose problems for the future. Exploration wells are already costing up to £5 M each, and individual projects over £1000 M. As smaller structures in deeper waters are developed, costs may treble in real terms over the next 10 years. The biggest projects may be in the £10000 M range. Such investment levels, which may be estimated in the order of £200000 M in the next ten years world-wide, are only within the capability and scope of international oil companies with the technology, management and asset base to support such activities. The consequences of such investments cannot yet be gauged, although it is likely that long-term rates of economic growth may be constrained by an altogether higher energy-cost plateau.

Since at least half of the offshore prospects may reside in O.P.E.C. countries and less developed countries (l.d.cs), this is of particular significance. First, during the past decade the position and importance of state oil companies within these areas has become increasingly predominant. Consequently western oil companies now often find themselves restricted to supplying management and technical services on a risk basis; that is, supplying the capital for exploration and development to be recouped only in the event of success, and frequently with rigid limits on the rates of return which may be expected. The temptation to move into barter deals, where imported goods and services are paid for by oil and gas supplies, produces further complications since such arrangements may involve both government-to-government and inter-governmental department negotiations. Secondly, there is a marked reluctance by private investors to embark on projects in some Third World countries because of the risk of political expropriation of their assets without adequate recompense. This is a particular hazard in the case of energy development projects and it is here that the question is raised whether there can be developed some form of guarantee, for example in conjunction with the World Bank as recently suggested by President Carter, that will ensure that the much needed investment to develop the potential offshore reserves of l.d.cs will not only be forthcoming but will be adequately protected.

Dominant is the rôle of the U.S.A., which in 1976 consumed 35 % of the free world's supply of oil with only 8 % of the proven reserve, and also consumes 62 % of the free-world gas sales while having no more than 15 % of the proven gas reserves. The rapidly growing dependence on imported oil on the one hand and, on the other, the need to conserve, economize and develop alternative energy sources will dominate energy policies as reflected by President Carter's recent statements.

The pattern of availability of U.K. offshore oil and gas supplies is markedly similar to the pattern of supplies for the free world, if our forecasts for new discoveries in the U.K. sector of the North Sea are correct. The experience gained should be directed with confidence to the world-wide opportunities, where there is a foreseeable need to increase the non-Communist world's oil production from some 50 to 65 or 70×10^6 barrels per day in the next 10 years, a large proportion of which increase should come from non-O.P.E.C. areas, and significantly from offshore areas. This (together with ambitious programmes for other types of energy) is a challenge to all in the industry if economic growth in the western world in the next quarter of a century is to be kept in the range of 2–3 % per annum.

Discussion

G. L. HARGREAVES (*Petroleum Engineering Division, Department of Energy, Thames House South, Millbank, London SW1P 4QJ*). Dr Birks has mentioned the very severe environmental loads to which fixed production platforms are exposed and further stated that such structures are very sensitive to deck loads, which must therefore be kept to a minimum.

Has Dr Birks considered the possibility of placing only a terminal structure above the sea's surface, connected by a shaft, or shafts, to production facilities housed deep down on the seabed, all but immune to wind and weather and where weight would be a stabilizing factor rather than a disadvantage. How would such an installation affect the economic viability of marginal deep-water oilfields?

J. BIRKS. I should like to consider Mr Hargreaves' question in the context of transferring the major production facilities to the seabed rather than the means by which this might be achieved. If we consider only the gas-separation process then the problems posed are by no means insurmountable provided the system can ultimately be tied back to the surface. Such a subsea separation system was in fact installed offshore Abu Dhabi in 1970 and commissioned in 1971 for a joint BP/C.F.P. development programme. Although at that time problems were experienced in the control and electrical systems, technology has subsequently been developed where such a system can be considered to be entirely feasible. However, when we consider the total production system requirements for a field development we must also cater for water process/injection, gas process/injection and power generation, etc. These requirements result in a very significant power load, normally at least 30 MW, and if these are enclosed in a seabed chamber or shaft the problem of maintaining a safe environment would be formidable. Earlier experience by BP with unmanned production facilities has not been very encouraging, both from a safety and efficiency point of view, and therefore the problem of operating personnel at the seabed would also have to be contended with. While we in BP have considered such concepts superficially to identify problem areas we cannot claim to have studied them in great detail for the reason that the present developments referred to in my paper appear to us to be more viable and practical solutions for water depths of up to 650 m.

I should perhaps also stress that the benefits of weight reduction are not entirely related to the technical feasibility of the systems but more importantly to extend the viability of such developments. I would suggest that when serious development is applied to total subsea production systems the volume/space requirements for such equipment will be of equal importance to the viability of such concepts. With regard to Mr Hargreaves' final question related to the effect of such total subsea systems on marginal field developments I can only reply that it is my present view that they would be less favourable than the concepts to which I have referred in my paper.

E. G. WEST (49 *Oxford Road, Stone, Nr Aylesbury, Bucks. HP17 8PD*). The excellent review by Dr Birks demonstrates very clearly the interrelation between the many technical and economic factors which tend to increase by an order of magnitude as the greater depths of the water are explored and exploited. Of the many comments and questions which come to mind, I will mention only two. The economic factors are inevitably influenced by political questions and an area of special importance is the possibility of enhancing the recovery of oil, which involves some quite complex scientific and technological problems. I feel sure that we should all appreciate information on any measures which could be used, or on which further research should be under-taken, to improve the yield from existing fields which, we understand, is considerably below 50 %, and averages only 35 %. Admittedly the cost of enhancement measures is high and it may well be that conventional means are uneconomic today but as shortages develop and fuel prices rise or as political–economic factors change, it is likely that the cost of additional recovery would be worth while. Oil left in the reservoirs now might not be recoverable after a lapse of time and this suggests that means to increase the yield should be in operation without delay, otherwise more than half the oil in the existing fields will be lost while we are recovering oil from newer wells in deeper waters at still higher cost. Hence, further research into enhancement should be undertaken sooner rather than later and the full economic factors need to be set out. Are there plans in hand for work on the necessary scale? Surely it should be on an international basis as a major contribution to the world energy problem.

An entirely different topic arises from my interest in structures. It was noteworthy that there is the need to reduce the weight of the superstructures, accommodation modules, etc., and this suggests that attention should be turned to the structural use of aluminium alloys for such items. Taking full account of the different physical and mechanical properties of aluminium alloys and steel, there can be overall weight savings of 50 %. The value of such weight saving can be set against the additional cost of aluminium over that of steel. The cost of fabricating need not be different and aluminium alloys for marine conditions can be readily welded. There would be further advantages in reduced handling and transportation costs of complete structures and of structural elements. In addition there are substantial longer-term savings due to the better corrosion resistance of aluminium compared with steel, leading to less frequent and lower expenditure on maintenance.

J. BIRKS. In reply to Dr West's comments, I would first emphasize that oil companies have for many years been seeking to improve the percentage of oil recovered and have carried out extensive research to this end. The relative lack of success, except for the now well-established technique of water injection, is itself an indication of the technical difficulties involved. While it is possible to achieve 100 % recovery on a rock sample in the laboratory by chemical methods, any improvement on a field scale is a much more complex problem with each field presenting its own combination of difficulties.

While there is some technical debate at present on the relative merits of early and late application of enhanced recovery techniques, there is of course a strong economic incentive to achieve any improved recovery during the normal life of the field facilities, particularly offshore. On the other hand, it would be unrealistic in present circumstances to consider delaying development of new fields in the hope that better techniques will become available quickly. None of the currently available techniques appears to be economic.

BP recently hosted a symposium on enhanced oil recovery, by which the experiences of major Oil Companies and Research Institutes in this field could be exchanged. The majority of the field trials on enhanced recovery were conducted in the United States and invariably the nature and homogeneity of the oil-bearing intervals resulted in inconclusive results on the effects of additives on water floods. Another practical problem is the high degree of absorption of such chemical additives by the formation rocks themselves, resulting in rapid dilution of the active agent at the water/oil interfaces. Emphasis was directed in this symposium to sea-water injection operations, which are likely to be the predominant recovery processes in the North Sea oilfields.

In this the most significant and economic impact of additives could be the improvement of injectivity rather than significant improvement in oil displacement efficiency. BP are planning to initiate field trials in the Forties field of such additives later this year.

The consideration of future oil shortages has led various governments to support more research in this area. In the U.S.A. a major programme is being supported by the Energy Research and Development Authority to the extent of $60 M p.a.; in Europe the E.E.C. is already supporting research in its member countries apart from that supported by the national governments. In practice, the results of this work become generally available through publications and conferences so that the effort may be regarded as a truly international one. As the research moves more from the laboratory to the field experiment stage, the costs and the technical uncertainties will rise steeply and the scale of the support may have to rise correspondingly.

The possibility of reducing platform superstructure weight by replacing steel with alternative lighter materials has been an active interest for a number of years, but the progress made in actually using such materials has so far been rather limited. The various materials that might be used all have some disadvantage, and in the case of aluminium alloys the main problems have been sparking hazard, loss of strength in the event of fire, and the difficulty of ensuring satisfactory standards of workmanship in using a relatively unfamiliar material. The main present use of aluminium alloy is probably in the fabrication of accommodation units. There will undoubtedly be an increasing use of this and other alternatives to steel, but the overall room for reducing the weight of platform superstructures by this means seems likely to be rather marginal.

Phil. Trans. R. Soc. Lond. A. **290**, 21–31 (1978)   [21]

Printed in Great Britain

The law of the sea and the deep seabed

BY SIR ROGER JACKLING†

Formerly of H.M. Diplomatic Service

It is the object of the third United Nations Conference on the Law of the Sea to obtain broad international agreement on the limits to the territorial sea, on that area beyond these limits within which the coastal state may exercise rights over living and non-living resources and on the nature and manner of exercise of those rights. The Conference is also required to establish an international régime to deal with the exploration and exploitation of the deep seabed beyond the limits of coastal states' rights. The work done by the Conference in five sessions since 1973 will have its effect on international law and practice but, partly owing to differences between the viewpoints of less industrialized and the more industrialized states (not confined to marine matters), the global solution essential for the orderly regulation of movement of shipping, scientific research and development of fisheries and sea-bed mineral resources may yet elude the Conference, to the detriment of the participating states and of the international community as a whole.

THE CONVENTIONS OF 1958

If until recently there was a broadly recognized body of international maritime law, it was enshrined in the four Conventions of 1958 – the Conventions on the Territorial Sea and Contiguous Zone, on the Continental Shelf, on the High Seas, and on Fishing and Conservation of the Living Resources of the High Seas. The product of the Conference of 1958, these conventions were in the main a codification of what could be said to be generally accepted internationally. But to a great extent this was a codification of what was acceptable to the traditional maritime powers. I do not need to delve into history to demonstrate that the concept of the freedom of the high seas, and the limitation of territorial waters to three nautical miles breadth was of primary interest to the powers with navies dominant on blue water. It is perhaps significant that the first breach in established concepts was made by the U.S. in the Truman Proclamation subjecting the resources of the continental shelf to its control, and announcing the intention to establish conservation zones for fishing. The significance to me is in the date: 28 September 1945, the time of the greatest relative power of the U.S. in military and naval terms. That these new ideas found their place in the 1958 Conventions is an important illustration of the part State practice plays in the development of International Law. But the many countries becoming independent since 1958 see them as freezing for all time a system evolved for the benefit of the established powers. That system has been challenged as discriminatory against the developing countries.

The Territorial Sea Convention had its value. It established the methods by which baselines were to be drawn, and incorporated a definition of islands. It provided for the right of innocent passage through territorial seas and stated the conditions to be observed. Provision was also made for a 12 nautical mile wide Contiguous Zone for certain customs and other similar functions.

† Present address: 37 Boundary Road, St John's Wood, London NW8 0JE.

But the Conference had failed to reach agreement on the breadth of the territorial sea, as did the 2nd Conference in 1960.

The Continental Shelf Convention gave formal recognition to the principle of the Truman Proclamation by granting sovereign rights to the coastal state for the purpose of exploration and exploitation of its resources, but its definition of the shelf was imprecise. The term was used as referring to the seabed and subsoil of the submarine areas adjacent to the coast but outside the area of the territorial sea to a depth of 200 m, or beyond that limit, to where the depth of the superjacent waters admits of the exploitation of the natural resources of the said areas. I do not have to point out in this company that the test of exploitability has rather different dimensions in 1977 from those envisaged in 1958.

The High Seas Convention codified the classic tradition: the high seas meant all parts of the sea beyond territorial waters and 'being open to all nations no state may validly purport to subject any part of them to its sovereignty'. The freedoms of navigation, fishing, overflight and cable-laying were confirmed, and warships on the high seas were stated to have complete immunity from the jurisdiction of any state other than the flag state. An obligation was laid on states to prevent pollution of the sea by discharge of oil from ships and from exploitation, and from the dumping of radioactive waste. The Fishing Convention has no direct relevance to the questions concerning this meeting. I make only two points: the Conference failed to get agreement on limits – the second attempt in 1960 failed by one vote – but did recognize a general obligation to cooperate in the conservation of the living resources of the High Seas. It also recognized the special interest of a coastal state in the maintenance of fishstocks in areas adjacent to its territorial waters, and a conditional right to adopt conservation measures unilaterally.

If the four Conventions of 1958 gave a formal status to traditional concepts they also gave some fairly clear pointers to the future. The Contiguous Zone concept, if its inclusion did something to blur the differences between the three-mile-limit nations and those claiming wider jurisdiction, was nevertheless for most states an extension of jurisdiction. The Continental Shelf Convention extended jurisdiction over the seabed without effective limitation. The High Seas Convention itself, although stated in its preamble to be 'generally declaratory of established principles of international law', recognized a general obligation in its anti-pollution provisions. And the Fisheries Convention, with its licence to a coastal state to take unilateral steps for conservation in the High Seas adjacent to its territorial sea, is a striking invitation to coastal state encroachment on traditional freedoms.

The pressure for review

However, the inadequacies of the 1958 Conventions swiftly became evident. In 1958 most states were content to claim a territorial sea of 6 nautical miles breadth or less. There were only 13 claimants to 12 miles. By the time the present Conference opened more than sixty states – by some counts eighty – or the majority of the greatly expanded U.N. membership, were claiming 12 miles or more. Fisheries claims to 200 nautical miles were more numerous and I have only to mention the problems that Britain has had with Iceland to remind you of how real these problems are.

The evolution of doctrine is well illustrated by the contrast between the case involving Britain and Norway in the International Court in 1951 and the Iceland case in 1974. In the first it was common ground that fishing rights were coterminous with the territorial seas. In the second the Court spoke of state practice revealing an 'increasing and widespread acceptance of the concept of preferential rights for coastal states'.

The steady encroachment of state practice on traditional doctrine of freedom of the seas, and the recognition which the Court appeared to give to this encroachment, may perhaps afford a clue to the long-drawn-out proceedings of the Conference of the Law of the Sea. If an increasing number of states had become impatient of such restrictions on unilateral extension of jurisdiction as the 1958 Conventions imposed, they could see no advantage in any rapid attempt to resolve only the questions left open by the earlier conference. When, following the famous speech by Ambassador Pardo in 1967, the establishment of a legal régime for the deep seabed began to be considered, pressure also developed for a complete review of international maritime law. The opponents of an attempt at a comprehensive treaty based their arguments in the main on the length of time the elaboration of such a treaty would take and the need for early action to settle the status of the deep seabed, which the Assembly in 1970 declared to be the common heritage of mankind.

PRELIMINARIES TO A NEW CONFERENCE

The resultant argument, played out in the Seabed Committee of the U.N., ended in an acceptance of the viewpoint of those who sought a comprehensive treaty. In my judgement, such a result to the argument was both inevitable and sensible, given the increasing uncertainties which divergent practices were fomenting.

In consequence the agenda of the Conference embraces all the questions of jurisdiction and resource development of the oceans. The story of its development through the meetings of the Seabed Committee is itself a long and intricate one, which time prevents me telling here, but in the course of it almost every political problem commonly complicating international affairs surfaced to plague us at one time or another.

Finally the Assembly called the Conference into being. Its first session was held in December 1973, at which the Committee structure was established and its officers elected. The problem of rules of procedure remained. The maritime states had in the main sought for a conference limited to the essential issues: to establish the area of 'the common heritage of mankind' and the régime for its exploitation. It was the developing countries who pressed for a comprehensive treaty. It had already become evident that many coastal states would seek extensions of jurisdiction which could gravely impair issues of major importance to the maritime countries. The concept of an Exclusive Economic Zone (E.E.Z.) of 200 miles breadth came in. Claims were being made which would hamper passage through straits, and threatened a plethora of uncoordinated regulations about pollution to the detriment of normal shipping movement. Scientific research was also at risk.

Because the maritime states with the most interests at stake – the U.S., the Soviet Union and her allies, the countries of the European Community and Japan – could muster relatively few votes by comparison with the countries of Africa, Latin America and mainland Asia, the Seabed Committee had from the start agreed that it would operate by consensus, and it continued to do so.

Clearly, however, an international Conference could not be committed to this principle. There had to be some provision for voting in the last resort. A formula was finally negotiated which should ensure that the majority cannot arbitrarily use its voting strength to impose unacceptable provisions on the powers with major maritime interests.

The second session of the Conference in Caracas ran for 10 weeks, the third, in Geneva in the spring of 1975, for 8 weeks. The fourth and fifth sessions were held in New York in the spring and summer of 1976, and the sixth session is at present (summer 1977) under way in that city.

PROCEDURAL PROBLEMS

The 1958 Conference had one great advantage over the present. It had before it from the outset draft texts prepared after years of work by the International Law Commission embodying the highest common factor of agreement as to the law then generally accepted. But inevitably the new Conference had to break new ground. No predigested single text was possible. So, with an exhortation by the Assembly to proceed by consensus the Conference set to work in Caracas on all extant proposals put to the Seabed Committee. This offered up to 16 variants of some basic matters, and the Conference churned slowly through this mass of paper without being able to establish anything in the nature of agreed positions. Trends, however, were beginning to emerge.

One other feature of the Conference must be mentioned. There were of course the usual geographical groupings of Africa, Asia, Latin America and Europe, and as has become customary at U.N. Conferences efforts were made by the geographical groups to establish defined positions on which they would remain firm. There were, however, conflicts of interests which made group decision on policy issues difficult, notably the difference between the landlocked states, who formed their own group, and the coastal states.

There had also developed, from early discussions between certain of the coastal states under the leadership of Norway, Canada and Australia, a grouping of states with divergent interests who sought to find a basis for agreed progress. Mr Evensen of Norway served as their chairman. Though the numbers of this group were at first limited, they reflected the principal interests of the Conference. The group's value lay in the informality of its proceedings. There were no records, and no formal secretariat participation. It was a carefully observed convention that every participant spoke in his personal capacity, thus preserving his right if need be to repudiate himself in public session. Already before Caracas very useful progress began to be made in this group in the beginnings of reconciliation of different positions. However, the very informality of the proceedings and the limitation of participants led to some suspicion and some jealousies. Gradually, as time went on and meetings of the group continued between sessions of the Conference proper, the number of participants expanded, and it was eventually agreed that although the meetings should be 'closed' so far as public attendance and records were concerned, any delegation had the right to sit in, and speak if desired – in a 'personal' capacity. The work of this group took on further importance as Mr Evensen on his own responsibility prepared and circulated suggested texts of issues central to the Conference on which he thought progress towards a compromise might be founded. After further discussion in the group Mr Evensen rewrote the texts on a number of instances, often more than once. His personal contribution has been invaluable.

When the third session of the Conference opened in Geneva in March 1975 there was some hope that enough had been done to permit real negotiation, either in the formal committee sessions or at least in informal meetings. But it was not to be, and the lack of progress compelled a reassessment of procedure. It was clear that the absence of anything like a basic text on which to negotiate was one of the troubles. Without such a focus it was too easy for those who had no taste for early agreement to obstruct progress.

After 4 weeks the proposal was therefore made and accepted by the Conference that the chairman of its three main Committees should each prepare a single negotiating text, which should take account of all discussions to date, would be informal in character, and be without prejudice to the position of any delegation. Three texts were duly prepared and circulated on the last day of the Conference. This was by design, since it was obvious that if the device was to serve a useful purpose Governments must have time to consider and to consult on these texts before Delegations took positions on them. As a result of further sessions the texts are now before the Conference in revised form.

PROGRESS TO DATE: THE RÉGIME FOR THE DEEP SEABED

How do matters stand now? This is the best seen by reference to each Committee in turn.

The first Committee of the Conference is concerned with the régime for the seabed beyond the limits of national jurisdiction. These limits will be established by the work of the second Committee, but the limits to national jurisdiction will at the least be at the 200-mile mark and may extend to the edge of the continental margin where that lies beyond the 200-mile mark. We are therefore talking of the deep seabed beyond the continental rise – and if I attempt no clearer definition of the margin at this or indeed at a later stage there are many who will understand and sympathize.

The Conference has to devise a method by which the resources of this area can be developed, consistent with the Assembly declaration of the common heritage of mankind. Let us assume that the resources in question are the manganese nodules lying at or near the surface of the deep seabed at depths for the most part of 12 000 ft or more. There are of course questions still at issue regarding petroleum resources, at least until 'the limits of national jurisdiction' have been agreed, but let us leave that aside for the moment.

It is only in the last decade or so that exploitation at these great depths has appeared possible. For the industrial countries new sources of nickel and copper are needed. The developing countries are eager for new sources of wealth to promote development world wide. There is also concern on the part of present producers that new sources may damage markets.

It is common ground between these parties of differing interests that an International Authority shall be established to exercise surveillance over resource exploitation. The differences of course are over the powers to be exercised by that Authority and the manner in which decisions over the exercise of that power should be reached.

The Group of 77 wished to establish that the right of exploitation lay with the International Authority, and that it should be solely for the Authority to decide whether there should be any contracting out. The industrial countries were concerned to ensure access to and production of deep seabed minerals by states and their nationals under reasonable conditions with security of tenure.

Various means of bridging this difference of approach have been explored over the past 5 years. It has been a period during which the basis of resource exploitation all over the world has been undergoing radical change, and much closer control by national governments has become the rule. In this climate the concept of concession granting by the International

Authority on anything like the terms to which the international companies had been accustomed in the past was swiftly seen to be unnegotiable. But the degree of authority claimed by the 77 was inconsistent with the sort of assured tenure required to justify the enormous cost of prospecting and mining.

Nevertheless, some progress has been made. If agreement is reached on a text it will, I think, broadly follow the shape of the Revised Single Negotiating Text (S.N.T.) – although further changes of substance would be necessary to achieve a compromise. This part of the text runs to 77 pages, with 63 articles and 3 long annexes, and I can give no more than the most impressionistic of descriptions of it. It is, however, a conference paper in the public demesne, and obtainable from the usual sources.

The first 19 articles are concerned with the definition of the area and its limits. Of course the extent of national sovereignty over the seaward projection of the continental land mass is still in dispute, but is for settlement elsewhere in the Convention. The text provides that the area and its resources are the common heritage of mankind, that no state may claim or exercise sovereignty over any part of the area or its resources and that activities shall be carried out for the benefit of mankind as a whole, taking into particular consideration the interests and needs of the developing countries. Article 9 is of crucial importance in laying down the general principles regarding the economic aspects of activities in the area. This clause has been much expanded in the Revised Text. It requires these activities to be developed so as to increase the availability of resources to meet world demand, but elaborates on the protection of existing developing country producers by commodity agreements in which the International Authority should take part. It provides that total production shall not exceed the projected cumulative growth of the nickel market during an interim period of 20 years.

POWERS OF THE PROPOSED INTERNATIONAL AUTHORITY

There are provisions relating to scientific research, giving the Authority power to conduct research, but not attempting to give any exclusive right to the Authority. There are also provisions to encourage the transfer of technology to developing states, to protect the marine environment and regarding damage liability. In all of these there is much ground now common to all delegations.

Now for exploitation. The governing provision here is Article 22. This empowers the Authority to conduct activities directly. A later provision establishes an organ called 'the Enterprise' to be the operational arm of the Authority. States or enterprises either state or private may also operate in association with the Authority on a contractual basis. The terms for such contracts are set out in Annexe I. Of these provisions I would say only that the Authority would control the contractor at all stages of the operation to ensure compliance with the contract and all provisions of the convention. The contractual concept is in itself a compromise between the earlier insistence of the developing countries on the Enterprise as alone having the right to conduct mining operations and the wish of the industrial countries for a licensing system, to be administered by the Authority. In his report on the fifth session, the Chairman of the first Committee noted that developing countries now generally accept that, as well as the Enterprise, other entities such as companies may also participate in mining operations in a form of association with the Authority. The common interest, in encouraging rapid and efficient seabed mining operations in order to increase supplies of raw materials, does find recognition in the Revised Text, as does the demand for protection of land-based producers.

The system envisaged is in effect one of parallel access. The annexe also provides in paragraph 8 that the Authority shall enter into negotiations with an applicant who meets the stated requirements with a view to conclusion of a contract. This obligation on the Authority would seem especially important in the context of the requirement for assured access. The parallelism is enhanced by paragraph 8 (*d*), which the Committee Chairman elaborates in his covering note. He explains that in applying for a contract the applicant will specify an area twice as large as the intended mine site, or two areas of equivalent promise. If a contract is concluded, the Authority retains one of the two sites, which would then be available if it were so decided for direct exploitation by the operating arm of the Authority, to be styled the Enterprise.

Given the powers of the Authority in relation to the terms of contracts and their supervision, the organs of the Authority itself and their powers in relation to its day-to-day operations are of manifest importance.

INSTITUTIONS OF THE AUTHORITY

As to institutions, it is proposed that there should be an Assembly, a Council, a Tribunal and a Secretariat.

The Group of 77 have, as a reflection of their general approach to international problems – the Security Council would never have its present composition of permanent members if the U.N. charter were being written today – sought to insist on the absolute power of the Assembly. The Assembly is to meet annually, on a one-member one-vote basis, but with a provision requiring a two-thirds vote on questions of substance. The text proposes some interesting procedural devices reminiscent of those of the Conference itself, to reduce the risk of precipitate decisions. The executive organ of the Authority would be a 36 member Council, with the duty to ensure that the Authority acted consistently with the general policies to be prescribed by the Assembly.

The Council structure is intended to give reasonable representation to the various interests. Of the 36 members, 6 are intended to represent the industrialized powers directly, 6 from the developing countries selected to ensure representation of exporters, importers, the landlocked and so on respectively, and 24 in accordance with the principle of equitable geographical representation as understood at the U.N. Here a two-thirds 'plus one' majority rule is to prevail.

The Council is to arrange for the setting up of a Planning Commission, a Technical Commission and a Rules and Regulations Commission, all of which are advisory to the Council.

The relation between the Assembly and the Council is critical to the negotiations. The Chairman states in his report that a system based on the supremacy of one organ of the other 'could not constitute a compromise solution'. A system is therefore proposed under which the Council would have sufficient latitude to execute the various tasks assigned to it and to carry out day-to-day operations in accordance with general policies established by the Assembly.

Despite the efforts at compromise, there are obviously differences still to be bridged. The respective powers of Assembly and Council cannot in all probability be laid down in advance to the satisfaction of all parties. As the then leader of the U.S. delegation put it, 'any protection of industrial country interests built into the Council will be essentially nugatory if Council decisions may be revised or circumscribed by an Assembly operating on a one nation–one vote principle'.

Much work clearly remains to be done on this issue. Much work has also to be done in order to work out the respective rôles of the Enterprise and other entities and their relation to the Authority. Perhaps sufficient definition of these can be achieved to render differences of an ideological character about the supremacy of the Assembly appear of less moment. But agreement on these issues is vital to the outcome of the Conference as a whole.

THE CONTINENTAL SHELF

The work of the Second Committee embraces the most diverse of the issues before the Conference, covering the Territorial Sea and its limits, innocent passage, international straits, fishing rights, the extent of the continental shelf, high-seas rights, archipelagos, islands and related questions to all these matters.

While I shall concentrate on matters relating to the continental shelf and its extent, you will realize that the various provisions affecting the freedom of movement of shipping are of at least equal importance. The existing drafts are in the main satisfactory and provided they stick should prove an acceptable part of the package. If so the extension of the breadth of the territorial sea to 12 miles is acceptable to us.

How broad is the continental shelf? Under the present text the coastal state is entitled to establish beyond its territorial sea an economic zone extending to a maximum of 200 miles from the baselines from which the territorial sea is measured. In that zone the coastal state would have sovereign rights over living and non-living resources of the seabed and water column, and exclusive rights and jurisdiction regarding the establishment of artificial islands, installations and research. The fishing rights of the coastal state are subject to qualification, but the mineral rights are unqualified. For seabed resources, the 200-mile limit has to be considered in relation to the definition of the continental shelf, in Article 64, which reads:

> The Continental Shelf of a coastal state comprises the seabed and subsoil of the submarine areas that extend beyond its territorial sea throughout the natural prolongation of its land territory to the outer edge of the continental margin or to a distance of 200 nautical miles from the baselines where the outer edge of the continental margin does not extend up to that distance.

The next articles confirm the exclusive right of the coastal state to exploration and exploitation of the continental shelf.

This definition of the continental shelf is of great importance to the United Kingdom. As I remarked earlier, the test of exploitability, on which rights at depths beyond 200 m rested under the 1958 Convention, left matters somewhat unclear. For us the North Sea is not the only area of interest for its hydrocarbon potential. To the West of Scotland and in other areas around our coast there are geological structures which could well contain oil. Some of these areas off the West of Scotland lie well over 200 miles from the mainland. In September 1974 Her Majesty's Government (H.M.G.) designated under the Continental Shelf Act 1964 a considerable area of the shelf – although not all to which the U.K. is entitled. These rights are, in the view of H.M.G., enjoyed under existing international law, under the 1958 Convention and under customary international law as evidenced by state practice and enunciated by the International Court in the North Sea Continental Shelf cases. However, not all states share this view. A clear legal definition of the edge of the margin is necessary, therefore, if conflict is to be avoided, whether between states themselves or between coastal states and the International Authority. Such a proposal is now on the table although not in the S.N.T. The Chairman of the second Committee in his covering note indicated that there was significant support for the concept that the continental shelf extended to the edge of the margin and was therefore sympathetic to proposals for its precise definition. He thought, however, that the proposals put forward, because of their very technical nature, needed to be considered by a group of experts at the fifth session. No agreement was reached there, regrettably, and discussions are to be resumed at the present session.

Opposition to such a concept comes from many states who would not benefit, particularly the Group of Landlocked and Geographically Disadvantaged States. This Group includes such landlocked states as Austria, Nepal, Afghanistan and Switzerland and shelf-locked states such as East (and West) Germany who for geographical reasons can claim only a limited E.E.Z. or shelf. There are 57 states in this group and if they hold together they could block conclusion of a Convention if their needs were not met. But as a move towards compromise, the U.K. and some other coastal states have proposed that a share of the value of production (presumably mainly petroleum) in the area between the 200-mile mark and the outer edge of the margin should be given to a fund presumably for the benefit of developing countries.

RIGHTS OF PASSAGE

I should mention here one other matter of great importance to the Conference – that of the status of the waters of the economic zone. Do they or do they not retain that of the High Seas? From the point of view of the discussions here this question has perhaps little direct relevance, but the maintenance of freedoms of navigation, over-flight and cable laying and other traditional High Seas uses are of major importance to many users of the oceans. (The Chairman of the Second Committee takes the view, with which I have myself much sympathy, that the waters of the E.E.Z. are *sui generis* and in informal discussions there has been progress towards a compromise between those who claim the equivalent of territorial sea status for the whole E.E.Z. and those who regard as vital the maintenance of High Seas rights, other than those like fishing, quite incompatible with the E.E.Z. concept.)

Similarly the problem of vessel source pollution, while of central significance to the prospects of a Convention, has no direct relevance to our concerns here. However, on both these vital issues some compromise is in sight.

THE FUTURE OF RESEARCH

Of greater interest to this meeting is the position regarding marine scientific research. The key issue here is the extent to which research in the economic zone and on the continental shelf should be subject to the consent of the coastal state. Some states consider that prior consent of the coastal state is required before any research of any sort should be conducted. Others maintain that the need for consent should be confined to research concerned with the discovery and evaluation of economic resources. One can see the territorialists at work again.

Detailed control, exercised perhaps in different ways by neighbouring states, whenever planned within 200 miles of the coast or anywhere on the shelf, would be a serious hindrance to marine scientific research generally. Indeed, if prior and specific consent had to be obtained for voyages passing through the zones of a number of different countries, a research project could suffer interminable delays. Even a notification procedure would be hampering. A good deal more work is clearly needed on this.

Dispute settlement

I will make only the briefest mention of the fourth part of the Text, although its importance is obvious. This provides for Dispute Settlement Procedure, and envisages a Law of the Sea Tribunal. To many participating countries, notably the United States, the inclusion of a mechanism for the compulsory settlement of disputes is basic to the Convention. The attempt to exclude the E.E.Z. from this procedure, given the views I have described of some states that the 200-mile zone should be to all intents akin to a territorial sea, is not surprising. To quote the some time leader of the U.S. delegation, 'If states cannot resort to international adjudicatory procedures to protect their rights, they are ultimately faced with the same problems arising from unilateral treaty interpretation that arise from unilateral claims.' In other words, what would then be the point of the treaty?

Prospects

What then are the prospects for a Convention? Given the issues still unsettled, affecting access to the deep seabed minerals, institutions of the Authority, extent of coastal state jurisdiction, pollution, research and compulsory dispute settlement, a final consensus will require a package deal.

The determination of the U.S. administration to get a solution can perhaps be measured by the appointment of that experienced public servant Elliot Richardson, formerly Ambassador in London and a senior officer in a Republican administration, as leader of the U.S. delegation. This suggests a desire for a bipartisan approach, but, together with other changes in the Delegation, a fresh look at some of the issues. This, given a similar readiness on the part of 77, could make for faster movement. It has I think been true that the 77 have been reluctant to make any major change in their position during the meetings in the summer of 1976 because of hope that a Democratic administration might be more sympathetic to their claims.

On any commonsense assessment of remaining differences, the gaps would look fairly easy to bridge – certainly the Revised Single Texts now before the conference represent a great advance in themselves. But what I would term the theological differences, symptomatic of what we term nowadays the North–South conflict, still present obstacles. The recent failure of the Organization for Economic Cooperation and Development to agree on means for commodity price stabilization is not a good augury.

After the experience of the last three years I hesitate to suggest that if no conclusion is reached this summer, there will be no Convention. If the 'package' is agreed, there will still be work to be done before a Convention is ready for signature, and ratification and entry into force can of course take years more.

But if the Conference should finally break down, whether it were this year or next, its work to date would have made for considerable differences in what would be generally acceptable as maritime law. A 12-mile breadth to territorial waters would be respected universally. There would, I think, be no attempt to challenge the rights of the coastal states over the resources of the seabed and the superjacent waters up to 200 miles.

Those states who have yet to claim rights over resources up to 200 miles would be quick to do so, and what this Conference would have failed to establish would gain universal recognition through state practice. Pollution control might be more satisfactorily settled as a separate matter. The existing Intergovernmental Maritime Consultative Organization Convention of 1973, if ratified and enforced, would itself go a long way to meet the anti-pollution objectives of Committee Three as regards vessels. The degree to which unilateral powers of control could be exercised could, however, lead to serious conflict with consequent damage to international trade if left unsettled for long. Nor, so far as I can see, can the extension of state practice provide a framework for the discharge of rights and obligations of those who seek to exploit the deep-sea mineral resources. I would not, of course, attempt to assess the weight or lack of it to be given to the Assembly resolution declaring these resources to be the common heritage of mankind.

The loss of the opportunity to settle finally arguments about coastal states' rights to resources to the edge of the margin would be at least equally regrettable.

Perhaps the most dispiriting aspect of failure would be the demonstration it would give of the inability of the international community to mould a needed system of law by conference. It may be that such a failure would be no more than many had expected from so ambitious an attempt. But a successful conference would in my view have implications beyond even its subject-matter, and one must hope that the governments participating will be moved by the general as well as the particular considerations to ensure that a Convention of universal application can be concluded.

Phil. Trans. R. Soc. Lond. A. **290**, 33–42 (1978)   [33]

Printed in Great Britain

Hydrocarbon potential of deep water

By H. R. Warman

64 *Lancaster Avenue, Hadley Wood, Herts., U.K.*

[Plate 1]

In recognizing that we are at a very early stage of exploring the geology and hydrocarbon potential of the Earth's deeper water areas, an attempt is made to generalize the geological conditions that exist on the continental slopes, the rises and the remainder of the oceanic provinces.

An attempt is also made to relate the postulated geology of these provinces to the conditions of structure and stratigraphy required to yield commercially extractable hydrocarbons in what must be a high-cost operation.

It is fully appreciated that there is much that we do not know about the geology of the more deeply submerged regions and that no accurate costs can be attributed to methods of hydrocarbon extraction that have not yet been devised; nevertheless it is felt that the current stage of knowledge and feeling for the order of magnitude of costs is adequate to provide some indication of hydrocarbon potential. This potential is not considered adequate to give any optimism for the deeper waters providing substantial additions to the reserves of exploitable hydrocarbons.

1. Introduction

As a normal product of the diagenesis of organic material trapped in sediments, and subjected over the long periods of geological time to the temperatures and pressures of deep burial, hydrocarbons are everywhere a common constituent of the fluids of the upper part of the Earth's crust. The concentration of hydrocarbons into accumulations sufficiently large to be commercially extractable calls for the coincidental combination of various geological factors throughout long periods of geological time; hence large accumulations are rare and impossible to predict in any one unexplored basin. We are only just beginning the search for hydrocarbons in deeper water and therefore do not know the geological conditions obtaining in many of the deeper water areas. Any attempt to quantify the hydrocarbon potential of such an imperfectly known domain can only be an informed guess. To assess the recoverable hydrocarbons also requires consideration of the time and cost of extraction and again we lack enough experience of deep-water production systems to make anything other than 'broad-brush' guesses of their cost; this factor again militates against precision in any estimates of recoverable hydrocarbons.

It is impossible to cover the detailed geology of the large areas of deep water of the whole world; this paper can therefore only describe the generalities of the main types of deeper-water provinces and consider the premises and problems bearing on the likelihood of the existence of commercial hydrocarbons and guess as to their quantities.

From what is known it is considered that appreciable quantities of oil and gas will be found beneath the deeper waters and that extraction of these hydrocarbons will develop rapidly over the next two or three decades. It would seem, however, that the deep-water areas do not have the potential for producing as much oil and gas as the land and shallow-water areas of the world and that although the contribution from deep water will be significant it will not postpone to any great extent the relatively imminent shortages of oil to meet our current demands.

2. Requirements for commercial production offshore

Hydrocarbons can only be produced if the cost of their production can be commercially met either by price in the market or by their relative attraction in a planned economy. With other fossil fuels such as coal being in relative abundance, and with the availability of other forms of energy such as nuclear power, cost must in the long term and in any form of society be a critical factor in determining the use made of hydrocarbons. One of the relative advantages of oil in the last few decades has been its low cost. In the heyday of abundant cheap oil from the Middle East, small onshore oilfields in many parts of the world could not be developed commercially. Now that governments in the major producing countries have quintupled the cost of world oil by their increased tax take, relatively small oilfields onshore can be commercially developed, but even now in remote and undeveloped areas onshore fields of capacity less than about 50 000 barrels per day are not worth developing unless they are associated with other fields and production facilities.

Offshore fields of very modest sizes can be commercially exploited, but the inhibiting effect of steeply rising costs due to increasing water depths and to exposure to severe weather conditions can be illustrated by quoting some figures for the North Sea. In the British sector and under U.K. tax rules an acceptable commercial rate of return can be made from a field with recoverable reserves of around 40×10^6 barrels and individual wells flowing at a peak of 1000 barrels per day under as much as 50 m of water. A field of 10–20 million barrels of oil can be commercial, utilizing a floating production system that can handle few wells and modest production, but in this case well productivities need to be in the range of 5000–10 000 barrels per day. In the main oil-producing area of the North Sea, i.e. with water depths in the range of 100–200 m, to meet the same financial criteria it is necessary to have fields in the range of 200–300 million barrels of recoverable reserves and well productivities in the range of 3000–5000 barrels per day at peak and maintainable for several years. It must be remembered of course that northern North Sea conditions are unusually severe and that costs of bottom-resting platforms in offshore areas with less severe combinations of wave heights and wind strengths are considerably less.

If we attempt to project our assessment of profitabilities out into water depths significantly greater than 200 m we must do so largely on conjectured costs of engineering concepts backed by little or no experience. Such a projection would indicate that at today's prices and today's value of oil, fields with recoverable reserves of at least 0.5×10^9 barrels and productivities of 5000 barrels per well per day would be required. The incidence of such fields in the world is limited. To date the total number of fields discovered in the world with oil or a quantity of gas of the equivalent thermal value is only 332, of which some 240 are oilfields (Klemme 1976). Of this total, somewhere between 15 and 20 % have reservoirs incapable of productivities likely to be economic in the deep offshore. Half of the giant fields occur in two basins (West Siberia and the Persian Gulf) which appear to be unique in their concentration of large fields. Outside these two basins such concentrations of giant fields are unusual and such fields are sparse and widely scattered. The North Sea stands out as exceptional and has 6 % of the total number.

After considering briefly some of the geological requirements for the formation of the large and high-productivity fields necessary to make deep offshore drilling attractive, I shall consider the likelihood of finding these conditions in deep water.

Nearly all large fields have a considerable element of structure in their trapping mechanism – in which a moderate degree of deformation is required to form large simple traps. Coupled with this, adequate reservoir qualities, particularly permeability, must exist in reservoir rocks to allow not only accumulation in quantity but also to allow extraction at high rates.

A sufficient quantity of organic material must be trapped in the sediments and subjected to the temperature (variable with the heat flow of the region) and pressures resulting from burial in order to generate hydrocarbons and cause their migration into reservoirs.

The range of depth of burial for oil generation appears to be from about 1 to 3.5 km with an optimum for giant fields in the range 1.5–3 km. Gas is most abundant in the range 3.0–4.5 km. Gas exists below that to considerable depths, in fact at least to 10 km, but below 4.0 km there is normally a marked and continuous decrease in reservoir porosity and permeability.

Deformation to form traps must occur at the correct time to collect the hydrocarbons produced during the main period of migration.

All these requirements must be met for the formation of large and highly productive fields.

3. Geological régimes of deep water areas

The oceanic domain can be divided into three main categories of geological units: the mid-oceanic ridges, the abyssal plains or ocean basins, and the continental margins; the latter include microcontinents and foundered remnants or slivers of continental crust.

(a) Mid-oceanic ridges

The combined area of the oceanic ridges and their associated rises, some 118×10^6 km^2 (Menard & Smith 1966), constitutes 32.7 % of the total oceanic areas. The ridges consist of oceanic basalts with virtually no sedimentary cover; the rises on the flanks of the ridges have basalts covered by only a thin veneer of sediments varying from a few tens to a few hundreds of metres in thickness. These sediments are mostly pelagic, with no significant matrix permeabilities that could produce hydrocarbons at a significant rate.

The combination of meagre sediment thickness and adverse reservoir character enable us to dimiss totally the hydrocarbon potential of this régime.

(b) Abyssal plains or ocean basins

The ocean basins are large areas of low relief, including abyssal hills and archipelagic aprons, but typically consisting of featureless plains. The basins, which are everywhere under more than 4000 m of water, constitute 41.8 % of the total area of the oceans. In all the oceans there are scattered volcanic features, some of which rise to the surface of the waters, or nearly so.

Our scattered knowledge of these basins depends on the combination of reflection and magnetic profiles, dredged samples, and the sparse but invaluable boreholes of the Deep Sea Drilling Project. It is clear that these regions have a uniform basaltic floor and the veneer of sediments is, like that of the ridge and rise province, too thin and lacking in permeable rocks to be considered seriously as a major contributor of producible hydrocarbons. Evidence of

3-2

gas and of some oil should not be allowed to generate too much enthusiasm for the prospects of producible hydrocarbons. Details of the better known occurrences are given by McIver (1974) but his concluding paragraph comments '...each is an encouraging sign that one of the essentials for petroleum occurrence (i.e. source rocks) will be found in sediments presently under deep water. If there are reservoir rocks, if there are large enough traps, and if the source sediments have undergone enough alteration, large deep-water oil fields may be counted on one day...'. His optimistic conclusion could in my opinion be more suitably expressed as outlining the improbability of significant oil in the two régimes so far discussed which constitute some 75 % of the oceans.

One environment in the deep ocean areas that has potential reservoir rocks is that associated with coral reefs. Many of the volcanic and other topographic features that rise from ocean depths to near sea level have thick coral caps and flanking aprons of coral detritus. Many of these reefs and detrital aprons have been forming at least through much of the Tertiary and appear to obtain thicknesses up to 2000 m. Some of the aprons of detritus descend at steep angles to considerable depths; doubtless some guyots have coral caps but it is very unlikely that many, if any, of these young reefal developments are in contact with enough source material to have been filled with hydrocarbons, even in the rare cases where impermeable caps have been formed, and where burial of whatever source material there is has been adequate to generate hydrocarbons. Certainly significant numbers of oilfields are unlikely in these young and largely exposed reefs; in the rare cases where traps exist, gas is more of a possibility. In the island archipelagos and shallow seas of the East Indies the main hydrocarbon content of the buried Tertiary coral reefs and coralliferous limestones appears to be gas rather than oil, as for example in the Gulf of Papua, the southern South China Sea, southern Celebes. Oil is nevertheless present in significant quantities in reefs in western New Guinea and the adjacent Arafura Sea.

(c) The continental margins

The continental margins, as defined below, include the main deep-water areas that have significant prospects of producing hydrocarbons in significant quantities. Oil and gas have already been proved in the shallower parts of this environment and similar geology can be traced on seismic profiles into the deeper water where similar prospects for hydrocarbons must also exist. The main uncertainties in assessing the oil prospects are related to those parts of the margins where the geology is not that of the continental shelf, foundered or declined under deeper water, but in large volumes of sediment deposited in deep water and having always remained under deep water and in the geological environment peculiar to those regions. Representatives of such deeper-water deposits are known from the examples which have been lifted up and are now known as exposures on land or from drilling in shallow waters.

The continental margin as considered here is from the shelf break at or around 200 m, down the slope and out to the edge of the land-derived sediments which form a prism of variable thickness and width and which are more or less synonymous with the physiographic feature commonly called the continental rise.

In a paper of this scope it is clearly not possible to give detailed descriptions of the many and varied margins but some generalities of the main types of margins will be sketched and some examples considered briefly.

Although there is much variety in detail the two main types of continental margin are distinct and have very different prospects for hydrocarbons. The passive or pull-apart types characterised by the margins of the Atlantic are also present around the Indian Ocean, the Arctic Ocean and around Antarctica. For reasons that will be discussed later the Atlantic type margins have better prospects of containing exploitable hydrocarbons than the collision margins of Pacific type, typical of most of the deep water margins of the Pacific Ocean.

4. THE MARGINS OF THE ATLANTIC OCEAN

Figure 1 is a simplified map (after Roberts & Caston 1965) which shows the total extent of sediments that could be remotely likely to contain extractable hydrocarbons, i.e. up to 1 km thick. A depth of 1 km is probably too little for the generation of oil, and the biochemical gas in shallow sands at less than this depth is not likely to be an attractive commercial prospect. The probable practical limit for any significant hydrocarbons is probably nearer to a minimum isopach in the 3–4 km range. In any case the thinner, oceanward, edge of these land derived clastics is almost certainly composed of extreme distal turbidite facies with little sand and high clay content, and hence holds little prospect of reservoirs.

Within the parts of the margins considered under our definition as being prospective there are two main categories of prospects. The first is in sediments that were deposited in shallow waters but which have foundered and now occur beneath the continental slopes or rises, and perhaps even in occasional segments of deeper oceanic floors which appear to be continental in origin.

(a) Shallow water sediments of Atlantic slopes

In most of the lands bordering the Atlantic there are thick sequences of Palaeozoic sediments; most of them have undergone various periods of disastrophism causing deformation varying from gentle folding to intensive metamorphism. Nowhere in these Palaeozoic onshore sediments bordering the Atlantic are there any oil or gas fields of the size and productivity required for commercial development in the deep offshore. There are no grounds for expecting any different conditions offshore and it is therefore considered unlikely that there will be any appreciable quantity of extractable Palaeozoic oil on the Atlantic margins.

Before the main drifting apart of the Atlantic margins there were developed extensive subsidiary basins in which were deposited considerable thicknesses of terrestrial and shallow marine sediments. Seismic profiles continuing out from the shelves allow confident identification of such sedimentation in many areas. In the early stages of rifting shallow water sediments probably deposited on the young oceanic basaltic crust. Roberts & Caston (1975) give a brief and clear discussion of the evidence for this. Evaporites are widespread along the margins of both the North and South Atlantic and may well cap underlying shallow water sediments containing suitable reservoirs. Diapiric structures, principally of halites, are locally abundant. The continuation of the oil provinces of Gabon and Angola from the shelf to deep water are discussed and clearly illustrated by Beck & Lehner (1974).

Reflexion profiles across typical slopes of oceanic margins (figure 2, plate 1) show the style of tectonics that typify many of these margins, i.e. with relatively little compressional or other folding, but down-to-the-ocean faulting and common diapiric intrusions.

Although oil and gas must occur in appreciable quantities in these sediments on the slopes there are great uncertainties in any attempts to envisage how much will meet the above-mentioned criteria for commercial extraction. There is no reason to believe that the hydrocarbon prospects are any better than on the adjacent shelf areas; the reverse is probably true as structures tend to decrease in number below the shelf edges and the regional oceanwards dip and down-faulting will in a general sense reduce closure. While exploration in many of the shelf areas around the Atlantic has been inhibited or prevented for reasons of politics or policy, a considerable number of wells have been drilled in shelf areas within the past decade or so and only a few fields have been discovered that approach our economic criteria, other

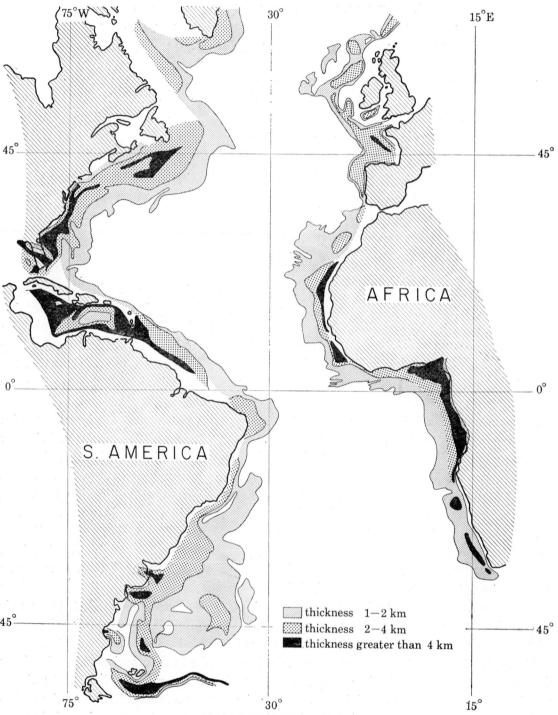

FIGURE 1. Atlantic sediment thicknesses.

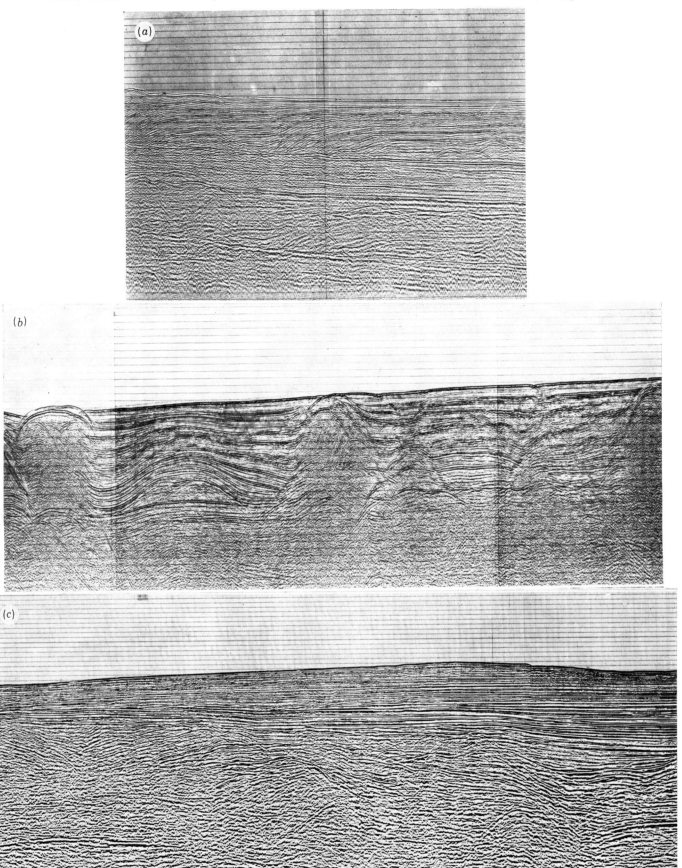

FIGURE 2. (*a*) Gravity tectonics above plane of décollement formed by messinian salt: W Mediterranean. By courtesy of the Western Geophysical Company. (*b*) Halokinesis, including classic 'turtle-back': Angola. By courtesy of J. Sefel and associates. (*c*) Pre-rift structurally deformed sediments underlying break-up unconformity: NW Australia. By courtesy of Geophysical Service Incorporated.

than those of the rather unique depositional and tectonic environment of the North Sea. Even with an increased tempo of exploration due to the increase in value of oil and the increase in availability of drilling equipment for the more exposed offshore areas it is difficult to imagine more than a further ten such oilfields and perhaps a similar number of gas fields being found in the next 10 years. As a guess one could attribute to these some $7\text{--}10 \times 10^9$ barrels of oil and $15\text{--}20 \times 10^{12}$ ft^3 of gas, these figures being of recoverable quantities.

(b) Deep water sediments of Atlantic slopes

An important domain in which hydrocarbon prospects exist in all oceanic areas but more particularly in passive, or pull-apart, types is in those sediments which have been laid down in the deeper waters and are by their nature related to the special depositional conditions below the edges of the continental shelves. The special conditions of gravity flow and turbidity current deposition are related to the gradients of the slopes and rises. Some examples of deposit of this type are known from onshore exposures where a combination of tectonic events and uplift have elevated such deposits above sea level. There are some examples of major oilfields in turbidite sands, including fields in Californian basins and some North Sea Palaeocene oil-fields such as Forties. However, these fields are in fact in turbidite sands of rather special character, which have accumulated within enclosed or semi-enclosed basins close to a source of quartzose sand.

All of the available evidence suggests that the predominant sediments of the Atlantic slopes and rises deposited in deep water, which are predominantly of Cretaceous and Tertiary age, consist of hemipelagic fine-grained material with at best only thin turbidite sandstones which are probably not only thin but also poorly sorted and with very little reservoir potential. The only deep-water sediments with much promise for hydrocarbons are the relatively localized concentrations of sands that occur as fans and slumped masses associated with discharge from submarine canyons, together with the spill of deltas that have prograded to the edge of the shelf, and by combinations of longshore currents, mass sediment flow and turbidity currents have provided great thicknesses of coarser clastics. Although the conglomerates and massive sands of some of these bottom-of-the-slope deposits can and do provide excellent reservoirs, they tend to be very localized in channels and fans and pass rapidly into turbidite sands and silts of poor quality. While the presence of channels and fans can be detected on reflexion profiles, and doubtless with time and experience this ability will improve, traps will be un-predictable. Within the complex of channels and fans there are normally adequate hemipelagics to provide cap-rocks but there is little compressional folding along most Atlantic type margins, and structures tend to be gentle. As there is ubiquitous down-slope dip it only requires one thin poor sand to connect with the upper part of a sand body to provide an up-dip leak; this appears to be a problem in Palaeocene reservoirs in the North Sea. Some sand bodies must have adequate trapping by passage up-dip into impermeable facies but the risk factor will be high; this in an environment of expensive deep water exploration drilling.

There are, however, known exceptions to the lack of structures. Beck & Lehner (1974) and Lehner & De Ruiter (1976) describe and illustrate the large structures associated with the delta toe overthrusts in the deep water at the foot of the Niger Delta which must encourage the expectation that the oil-rich province of the Niger extends into deep water.

Organic material is commonly trapped in turbidites and although the facies does not always appear to be well endowed with good hydrocarbon source material the very high oil content per unit volume of sediment in the Los Angeles basin attests to the effective source and generation ability on occasions. Samples from the Deep Sea Drilling Project illustrate the apparent maturity of hydrocarbon genesis at deeper levels with hydrocarbons of increasing carbon number with depth. Although the low-temperature gradient in most young slope-and-rise deposits militate against hydrocarbon maturity at shallow depth, the thickness of several kilometres in many areas will ensure both oil and gas generation in adequate quantities.

Some of the more comprehensive papers on the mode of deposition of the sediments of the slope and rises are those by Emery (1969), Middleton & Hampton (1973), Sangree *et al.* (1976), Walker & Normark (1976) and Woodbury, Spotts & Akers (1976). These papers have comprehensive lists of references.

We can be confident that these deep-water sediments contain oil and gas accumulations, some of which will be large enough to justify the cost of deep-water exploration and production. No meaningful way is available to quantify the distribution of such fields. As a complete guess one could imagine perhaps ten oil 'giants' (recoverable reserves of 0.5×10^9 barrels or more) and a similar number of gas 'giants' (recoverable reserves of 3.5×10^{12} ft^3 or more). In such fields in the Atlantic one could therefore postulate perhaps 7.5×10^9 barrels of recoverable oil and a comparable thermal equivalent of gas. In addition, one can expect there to be in this geological environment a large number of smaller accumulations totalling perhaps two or three times the reserves in the larger fields.

5. Margins of the Indian Ocean

Prospects around the margin of the Indian Ocean are similar to those of the Atlantic with the considerable exception of the collision-type margins of the island arcs extending from Burma through the Andamans, Nicobars and the islands of Indonesia to Timor.

Along the coasts of Africa and India, particularly the former, terrestrial Triassic–Jurassic deposits such as the Karroo and later Mesozoic shelf deposits dip under thick prisms of deep water late Cretaceous and Tertiary sediments. With the exception of the oil sands of Madagascar a modest but significant amount of exploration along the east coast of Africa has been singularly unsuccessful in finding hydrocarbons in quantity or commercial significance. Although one cannot dismiss the possibility of hydrocarbons being found in the older sediments, it is difficult to attribute them with any quantifiable potential. An exception to this is on the Australian margin of the Indian Ocean, where the Permian to Jurassic clastics which have yielded some oil and a lot of gas drop along down-faulted margins into deep water. Particularly off the NW shelf of Australia large structures exist in this environment which have good potential for hydrocarbons, although with the main likelihood being of gas.

Deep water sediments, including gravity flow and turbidites of Cretaceous and Tertiary age similar to those described in the Atlantic, occur around much of the margin of the Indian Ocean and again these must locally have prospects. Considerable attention has been focused on the enormous sediment cones off the mouths of the Indus and Ganges, but on the basis of exploration to date it is suspected that these will in large part be very short of adequate reservoir sands.

It is again virtually impossible to quantify the reserve expectations of the deep-water regions of the Indian Ocean, but as a wild guess one is tempted to suggest figures similar to those postulated for the Atlantic.

For a fuller definition of the main prospective areas of the Indian Ocean reference is made to Schott, Branson & Turpie (1975).

6. MARGINS OF THE PACIFIC OCEAN

One cannot do justice to the extensive and complex margins of much of the Pacific in this paper but a few generalities can be made. Most of the typical collision margins of the Pacific have no significant deep-water prospects in beds older than Middle Cretaceous; such older beds have generally been too tectonized and metamorphosed to be prospective. Off the narrow shelves of much of the Pacific the descent to great depths of water is rapid. Prisms of Cretaceous to Recent sediment, often of great thickness, occur along the Pacific margins, but in these sediments off the classic arcs stretching from New Guinea to Alaska there is a high proportion of volcanogenic sediments, a factor that will downgrade the already poor reservoir prospects in the deep-water sediments. In some areas, particularly off coasts with good granitic sources for arenaceous sediments, reservoirs will exist with oil accumulations comparable to those of offshore California, but they will take time to identify and even then prolific fields for economic production will be hard to find.

Quantification of the hydrocarbon potential of the deep-water margins of the Pacific is, if that is possible, even more uncertain than for the Atlantic and Indian Oceans. It is, however, unlikely that reserves will exceed those predicted for either of these oceans.

Although there are parts of the Arctic Ocean margins that have some prospects similar to those of the other oceans, the problems of ice in deep-water drilling production make such prospects of no interest for the forseeable future.

7. HYDRATES

Although there are very large total volumes of hydrates dispersed through the top layers of sediments of the deep oceans it is at present inconceivable that these could gainfully be liberated and collected on any meaningful scale and they will not be further considered.

8. CONCLUSIONS

From this very cursory consideration of prospects for deep water producible hydrocarbons it can be said that the large areas of the deep ocean basins have little prospect of containing large hydrocarbon reserves. The continental margins down to the rise clearly do have prospects of producing both gas and oil but the incidence of fields of sufficient size and productivity to warrant production will be limited. Finding such accumulations will not be easy and the cost of exploration alone will be considerable.

Over the next decade or two it is difficult to visualize the finding of more than 20 or 30×10^9 barrels of recoverable oil and within that time span not a great deal of this discovered oil is likely to be on production. Gas in comparable quantities will probably also be found but the increased problems of getting it ashore means that little, if any, will be produced in water depths greater than 500 m before the end of the century.

Although it is the province of the engineering contributions to this symposium to predict the availability of systems for really deep-water production it is considered by many workers that we shall be restricted within the next couple of decades to production from the upper portions of the slopes – perhaps down to 2000 m. The prospects of oil or gas production from the lower parts of the slopes and from the rises would seem more likely for the next century rather than before the year 2000.

The above estimates and predictions are based on so much uncertainty and assumption that they are prone to gross error. There is room for some major surprises in both the amount of hydrocarbon discovered and the ability to extract it at economic rates. It is not long ago that we were saying that 100 m of water was the likely limit of economic production. When significant finds are made in deep water, and they undoubtedly will be, there will be an enormous acceleration in engineering development to produce it. There is certainly enough prospect of deep-water hydrocarbons to encourage exploration for them in the face of the impending shortage from other areas, but this search will be neither cheap nor easy, and with the cost of single exploration wells running at £5–10 M each the pursuit of more subtle exploration plays will require considerable financial encouragement and attractive terms.

I wish to express my indebtedness to Dr V. Caston of British Petroleum for his guidance and assistance in preparing this paper.

References (Warman)

Beck, R. H. & Lehner, P. 1974 Oceans, new frontier in exploration. *Am. Ass. Petr. geol. Bull.* **58**, 376–395.

Emery, K. O. 1969 Continental rises and oil potential. *Oil & Gas J.* **67**, 231–243.

Klemme, H. D. 1976 World oil and gas reserves from analysis of giant fields and petroleum basins (provinces). In *U.N. Conference on Petroleum, Vienna.*

Lehner, P. & De Ruiter, P. A. D. 1976 Africa's Atlantic margin typified by string of basins. *Oil & Gas J.* **74**, 252–266.

McIver, R. D. 1974 Evidence of migrating liquid hydrocarbons in Deep Sea Drilling Project cores. *Am. Ass. Petr. geol. Bull.* **58**, 1263–1271.

Menard, H. W. & Smith, S. M. 1966 Hypsometry of ocean basin provinces. *J. geophys. Res.* **71**, 4305–4325.

Middleton, G. V. & Hampton, M. A. 1973 Sediment gravity flows: mechanics of flow and deposition. In *S.E.P.M. short course on turbidites and deep water sedimentation, course notes* (eds G. V. Middleton & A. H. Bouma), pp. 1–38.

Roberts, D. G. & Caston, V. N. D. 1975 Petroleum potential of the deep Atlantic Ocean. In *Proc. 9th World Petr. Congress*, vol. 2, pp. 281–298.

Sangree, J. B., Waylett, D. C., Frazier, D. E., Amery, G. B. & Fennessy, W. J. 1976 Recognition of continental slope seismic facies, offshore Texas–Louisiana. In *Beyond the shelf break: AAPG short course, course notes* (eds A. H. Bouma, G. J. Moore & J. M. Coleman), pp. F1–F54.

Schott, W., Branson, J. C. & Turpie, A. 1975 Petroleum potential of the deep-water regions of the Indian Ocean. *Proc. 9th World Petr. Congress*, vol. 2, pp. 319–335.

Walker, R. G. & Normark, N. R. 1976 Deep water sandstones. In *Sedimentary environments and hydrocarbons AAPG/NOGS short course, course notes* (ed. R. S. Saxena), pp. 129–217.

Woodbury, H. O., Spotts, J. H. & Akers, W. H. 1976 Gulf of Mexico slope sediments and sedimentation. P. C1–C28. In *Beyond the shelf break: AAPG short course, course notes* (eds A. H. Bouma, G. T. Moore & J. M. Coleman), pp. C1–C28.

Phil. Trans. R. Soc. Lond. A. **290**, 43–73 (1978)   [43]

Printed in Great Britain

Geochemistry of oceanic ferromanganese deposits

By S. E. Calvert

Institute of Oceanographic Sciences, Wormley, Godalming, Surrey GU8 5UB, U.K.

[Plate 1]

Deposits of mixed manganese and iron oxides, with high concentrations of minor metals, cover large areas of the deep-sea floor. They occur as *nodules*, over a very wide depth range, but most abundantly on the abyssal sea floor in water depths between 4 and 5 km, and as *unconsolidated sediments*, *rocks* and *crusts*, which are restricted to areas of the world-wide active ridge system. The two types of deposit have different chemical compositions and are the products of different precipitation and accretion mechanisms on the abyssal sea floor.

Ferromanganese nodules grow very slowly, they generally have Mn/Fe ratios greater than or equal to 1, and they contain high levels of Ni, Cu, Co, Ba, Pb, Zn and Mo. Regional, inter- and intra-ocean variations in composition are marked. Such variations can be easily mapped throughout the Pacific and are most plausibly explained by a combination of the variety of metal sources on the sea floor (including abyssal water, biological debris and perhaps volcanism), and by the different environmental conditions under which the nodules form. Some of the variability in the contents of Ni and Cu, two of the metals for which the nodules are considered a valuable resource, can be interpreted as a reflection of intensive metal recycling and diagenetic reaction at the sediment surface.

Iron- and manganese-rich sediments, rocks and crusts accumulate quite rapidly, they have highly variable Mn/Fe ratios, and they contain a different suite of minor metals, generally at lower concentrations, compared with ferromanganese nodules. Compositional variations in this case are on the scale of an individual specimen as well as regional, although clear trends are not evident at the present time. They are thought to form by precipitation from the reaction products of newly extruded basalt and seawater; isotopic evidence also indicates that there is significant adsorption of metals from seawater by the poorly crystalline oxides produced by this reaction.

1. Introduction

The ferromanganese deposits in the deep sea represent an important class of authigenic mineral deposits forming at the present time. They result from diverse chemical reactions on the sea floor and can in principle be used to provide information on physical, chemical and biological processes taking place in the ocean as a whole. Significant advances have recently been made in providing information on their geochemistry, an essential prerequisite for understanding their wide compositional variations, their modes of formation and the distribution of particular deposits of potential ore grade.

There are essentially two types of marine ferromanganese deposits, representing two different modes of accumulation of iron and manganese. The well-known *ferromanganese nodules* are widely distributed over extensive areas of sea floor, as well as in many shallow-water environments and in lakes. They have been studied extensively since their discovery during the *Challenger* expedition (Murray & Renard 1891) and recent reviews of various aspects of their distribution and composition have been provided by Skornyakova & Andrushchenko (1972),

Cronan (1974, 1976a), Margolis & Burns (1976) and Glasby (1977). Ferromanganese deposits of quite different character are found sporadically on or close to the mid-ocean ridge system. They occur as unconsolidated *sediments*, and as *crusts and rocks* of highly variable morphology and composition. Their discovery is more recent, and they are attracting a great deal of attention. Bonatti (1975) provides a useful review of their main features.

I shall discuss the geochemistries of the two classes of ferromanganese deposits in separate sections, stressing the more recent data available at the time of writing, before attempting to compare and contrast their compositions and modes of formation. I shall not include discussion of the metalliferous sediments of the hot brine deeps of the Red Sea (Degens & Ross 1969) because of their unique compositions and modes of formation (Manheim 1974), requiring a separate treatment.

2. FERROMANGANESE NODULES

2.1. *General*

Ferromanganese nodules are generally spheroidal or discoidal concretions, most commonly a few centimetres in diameter, which are most abundant on the surfaces of unconsolidated sediments in abyssal water depths (4–5 km). Estimates of the surface concentration of nodules in the Pacific, from sampling and from sea floor photographs (figures 1 and 2, plate 1), yield mean values of 10 kg m^{-2}, but reaching 38 kg m^{-2} (Mero 1965). Nodules also occur sporadically within the sediments (Cronan & Tooms 1967). Ferromanganese crusts on exposed rock surfaces have compositions broadly similar to the concretionary forms and are conveniently included in the discussion of nodule geochemistry.

Ferromanganese nodules may or may not possess a central nucleus or core, consisting of rock or mineral fragments in various stages of alteration, skeletal material (carbonate tests, teeth or bones), fragments of older nodules, or pieces of sediment. The coating of ferromanganese and other materials is most commonly concentrically arranged around the nucleus, some individual layers and laminae being traceable around the entire circumference of the concretion (Sorem 1967; Sorem & Foster 1972). Individual shapes of nodules are often governed by the shapes of the nuclei.

2.2. *Chemical composition*

Ferromanganese nodules are composed of an intimate mixture of crystalline and amorphous phases. This includes oxyhydroxides of iron and manganese (see §2.3), detrital aluminosilicates, organic debris, mainly skeletal, and minor quantities of authigenic silicates, sulphates, etc. They are highly porous and contain a large proportion of bound water. The heterogeneous nature of the nodules must be borne in mind in any discussion of their bulk chemical composition.

Although the chemical compositions of sea-floor nodules from all major ocean basins have been reported, attention will be restricted here to the Pacific, where the data are most abundant. Mero (1965) and Cronan (1974, 1976a) have provided average composition for a few elements based on a large number of analyses. In view of the extreme compositional variability in the Pacific (see §2.5) and because most of the data consist of partial chemical analyses, selected individual analyses are shown in table 1, together with data on nearshore nodules and pelagic sediments.

Inspection of table 1 shows that Pacific nodules have Mn/Fe ratios close to or greater than 1 and detrital aluminosilicate contents, based on the Si and Al concentrations (Calvert & Price 1977a), of roughly 25 % by mass. Some significant fraction of the Fe, but not the Mn, in nodules

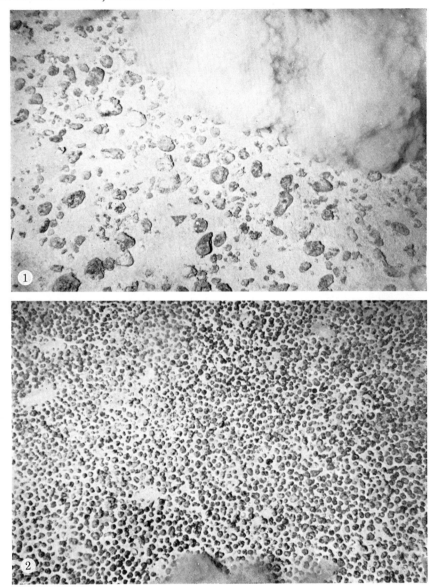

FIGURE 1. Sea-floor photograph at 20° 38′ N, 130° 46′ W, depth 5180 m. Nodules are 1–10 cm in diameter and irregular and discoidal in shape; many are partially buried in the fine clay substrate. Sediment cloud was caused by the camera touching the seabed. (U.S. Navy Photo.)

FIGURE 2. Sea-floor photograph at 13° 53′ S, 150° 35′ W, depth 3695 m. Nodules are 2–5 cm in diameter and are more or less spherical. The pale-coloured patches are mounds of sediment (calcareous ooze), probably thrown up by benthonic organisms, which cover the nodules. (U.S. Navy Photo.) The distance from the camera to the sea floor is larger in the bottom photograph.

is present in the aluminosilicate component so that the Mn/Fe ratio in the oxide fraction is higher than that shown by the data in table 1.

Of the remaining major elements listed in table 1, it seems clear that Ca, K, Mg, P and Ti are also present to some extent in the oxide fractions of the nodules. Calvert & Price (1977a) have shown that Ca, P and Ti contents are significantly correlated with the total Fe content, whereas Mg is related to the Mn content; K is probably located in the Mn-bearing phase.

TABLE 1. REPRESENTATIVE CHEMICAL COMPOSITIONS OF PACIFIC FERROMANGANESE NODULES, SHALLOW WATER NODULES AND PELAGIC SEDIMENTS

(Major elements in percentages by mass; minor elements in parts/10^6.)

element	nodules						sediments		
	1	2	3	4	5	6	7	8	9
Si	12.80	5.40	5.90	4.95	13.64	6.32	25.80	22.00	1.10
Al	2.60	2.00	2.40	1.80	4.35	2.27	9.20	6.50	0.40
Ti	0.62	0.56	1.27	0.89	0.08	0.21	0.65	0.38	0.12
Fe	11.40	6.40	16.40	14.05	1.36	3.92	5.60	4.50	0.80
Ca	1.60	1.60	2.20	6.85	0.87	5.56	0.70	1.60	32.10
Mg	0.90	1.70	1.40	—	—	1.87	2.00	2.00	1.70
K	0.80	1.00	0.75	0.39	1.52	1.03	2.85	2.07	0.05
P	0.15	0.19	0.33	0.08	—	0.35	0.08	0.40	0.22
Mn	13.20	24.90	16.00	14.4	24.8	30.20	0.80	0.91	0.60
CO_2	0.04	0.04	0.37	—	—	11.88	0.92	0.92	38.42
As	100	65	190	—	—	245	30	30	10
Ba	730	2420	1280	4300	3300	3090	690	5640	600
Co	2000	2400	4400	7000	170	120	190	280	290
Cu	2300	10100	900	720	460	17	300	700	200
Mo	250	610	375	430	320	55	40	40	15
Ni	4600	12500	2000	2900	1200	77	260	400	80
Pb	1080	560	1150	2000	460	42	90	45	100
Rb	25	15	7	—	—	40	205	90	5
Sr	630	530	820	1400	390	770	170	320	1190
Y	95	130	160	—	—	28	50	290	30
Zn	600	860	390	400	430	60	200	270	110
Zr	360	290	380	—	—	55	190	200	30

Sources: 1, Station JYN II 8G, 40° 29′ N, 172° 32′ E, 4250 m depth (Calvert & Price 1977a, Table I). 2, Station JYN V 47PG, 14° 37′ N, 135° 04′ W, 4813 m depth (Calvert & Price 1977a, Table I). 3, Station AMPH 85PG, 11° 35′ S, 158° 31′ W, 5338 m depth (Calvert & Price 1977a, Table I). 4, Station MP 33K, 17° 48′ N, 174° 22′ W, 1810–2290 m depth (Mero 1965, Table XXX). 5, Station VS BII-35, 22° 18′ N, 107° 48′ W, 3000 m depth (Mero 1965, Table XXX). 6, Loch Fyne, Argyllshire, 200 m depth (Calvert & Price 1970, Table 2). 7, Pelagic clay, Station JYN VI 11G, 27° 42′ N, 175° 10′ E, 5750 m depth (Calvert & Price 1977a, Table I). 8, Siliceous clay, Station JYN V 43PG, 9° 53′ N, 138° 56′ W, 4893 m depth (Calvert & Price 1977a, Table I). 9, Calcareous ooze, Station AMPH 80G, 11° 51′ S, 160° 51′ W, 3803 m depth (Calvert & Price 1977a, Table I).

The data in table 1 also serve to illustrate the large degrees of enrichment of a wide range of elements in ferromanganese nodules compared with pelagic sediments. The nodules display extreme fractionations for the first and second row transition metals and As, Ba, Sr and Pb, as indeed do red clays relative to nearshore sediments (Wedepohl 1960).

There is a large degree of variability in the chemical composition of Pacific nodules, first described by Mero (1962). Abyssal sea-floor nodules from the north and south central Pacific have similar compositions, while those in the northern tropical area are enriched in Mn, Ba, Mo, Ni and Zn (table 1). Nodules collected on central oceanic seamounts have roughly equal Mn and Fe contents and are enriched in Co and Pb. Nodules from the eastern marginal areas have very high Mn/Fe ratios and relatively low minor element contents, these features re-sembling nodules from some shallow-water areas of rapid sedimentation (Calvert & Price 1977b).

The relations between some of the minor transition metals and Mn and Fe in ferromanganese nodules have been examined using bulk chemical (Goldberg 1954; Riley & Sinhaseni 1958; Mero 1962; Willis & Ahrens 1962; Ahrens et al. 1967; Cronan & Tooms 1969; Margolis & Burns 1976; Glasby et al. 1974; Glasby 1976; Calvert & Price 1977a) and electron microprobe (Burns & Fuerstenau 1966; Cronan & Tooms 1968; Friedrich et al. 1969; Dunham & Glasby 1974; Ostwald & Frazer 1973; Lalou et al. 1973a; Burns & Brown 1972) analyses. Significant positive correlations have been observed by various authors between Ni, Cu, Zn, Mo, Ba, Mg, K and Mn and between Co, Pb, Ti, V, Mo, Ce, Zr and Fe. Burns & Fuerstenau (1966) proposed that simple substitution of cations into the Mn- and Fe-bearing phases (see §2.3) could explain the observed correlations and this has been broadly confirmed by later work. The importance of a knowledge of the mineralogy of the various oxyhydroxide phases thus becomes apparent.

The relation between Co and the major Mn and Fe phases does not appear to be simple, however. Burns (1965) and Burns & Fuerstenau (1966) suggested that Co$_{III}$ replaces Fe$_{III}$ in the iron oxyhydroxide phase in nodules and many bulk analyses of nodules show positive Co–Fe correlations. However, in some cases this relation is not observed, especially in abyssal nodules (Cronan & Tooms 1968; Ostwald & Frazer 1973). In addition, Price & Calvert (1970) pointed out that in seamount nodules the Co contents vary over very wide limits and do not appear to be related to the Fe content. For this reason, they suggested that Co$_{III}$ (Goldberg 1961; Sillén 1961) enters into the highly oxidized δ-MnO$_2$ phase as well as into the Fe oxyhydroxide phase. Burns (1976) has provided a plausible mechanism for this process, involving the substitution of low-spin Co$_{III}$ for Mn$_{IV}$ in the structure and van der Weijden (1976) has presented evidence for the presence of Co in both the Mn and the Fe phases in a suite of Pacific Ocean nodules from a range of environments.

A further instructive example of fractionation exhibited by ferromanganese nodules concerns the distribution of the lanthanides. Oceanic nodules commonly display a marked Ce enrichment and variable enrichments of the other elements, relative to shales, depending on water depth. Nodules shallower than 3–3.5 km depth are enriched in Yb and Lu relative to Sm, Eu and Tb, whereas deeper nodules show marked depletions in Yb and Lu (figure 3). The shallow nodules have lanthanide patterns (apart from Ce) resembling that of seawater (Høgdahl et al. 1968) which is used as evidence for the derivation of these elements in nodules from seawater by coprecipitation (Goldberg et al. 1963). By contrast, the lanthanide patterns for the deeper nodules appear to be the mirror image of that of seawater (Piper 1974a).

Note that unlike the remaining lanthanides, Ce can exist in the Ce$_{III}$ and Ce$_{IV}$ oxidation states in nature (Goldberg 1961). This could be the explanation of the extreme Ce enrichment in the highly oxidized nodules and the extreme Ce depletion in seawater relative to river water. In addition, seamount nodules, containing the more oxidized Mn phase (§2.3.1), have the most marked Ce anomaly (Piper 1974a), and indeed the highest concentrations of Co, Pb and Ti, metals which can also be oxidized in the marine environment (Sillén 1961; Price & Calvert 1970).

The depth at which the change in lanthanide patterns of ferromanganese nodules occurs in the Pacific (Piper 1974a) is close to the depth of the lysocline (Berger 1970), approximately 3.7 km where the solution rate of CaCO$_3$ shows a marked increase (Peterson 1966). Release of the lanthanide elements from dissolving biogenous tests and their coprecipitation with accreting ferromanganese oxyhydroxides could therefore explain the interesting contrast between deep and shallow-water nodules since Piper (1974a) has shown that foraminiferal tests, like the nodules, have Yb and Lu depletions relative to Sm and Eu.

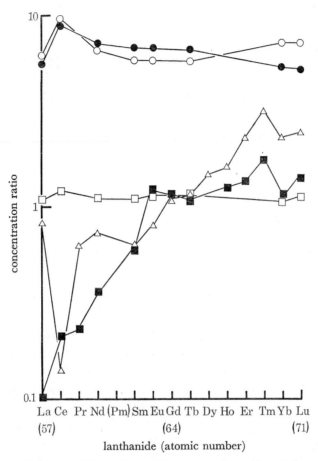

FIGURE 3. Distribution of the lanthanides in ferromanganese nodules and pelagic clay (from Piper 1974a), sea-water (from Høgdahl et al. 1968) and Mid-Atlantic Ridge basalt (from Frey & Haskin 1964); ●, nodules from below 3000 m; ○, nodules from above 3000 m; □, pelagic clay; △, seawater ($10^7 \times$ concentrations); ■, M.A.R. basalt. Concentration ratio is the ratio of each individual lanthanide to its concentration in average shale (Piper 1974a).

2.3. *Minerology*

Although the mineralogy of the manganese and iron oxyhydroxides in oceanic nodules has been difficult to elucidate, because of the very small sizes of the crystallites, the intimate inter-growth of authigenic phases and the presence of amorphous material, it seems clear from the available structural information that a great deal of the chemical variability of the nodules is due to a variable Mn phase mineralogy. There has been considerable confusion over nomen-clature in the literature and this has recently been reviewed by Burns & Burns (1977).

2.3.1. *Manganese phases*

Using X-ray diffraction techniques, Buser & Grütter (1956) identified two different crystal-line Mn phases in a small sample of oceanic ferromanganese nodules. They were referred to as $\delta\text{-}MnO_2$, described as an essentially disordered two-dimensional layer structure characterized by reflexions at *ca.* 245 and 140 pm, and a double-layered structure, similar to that of litho-phorite, consisting of MnO_2 sheets alternating with disordered layers containing Mn^{2+} and OH^- ions and H_2O. The structure of this latter phase was stated to be reasonably similar to that of the manganite group with the general formula $3MnO_2 \cdot Mn(OH)_2 \cdot xH_2O$ (Buser *et al.* 1954) having a characteristic basal reflexion at around 1000 pm. Grütter & Buser (1957) examined some other nodule samples and, in addition to the two phases already noted, identified a separate manganite phase, characterized by a reflexion at 710 pm. This phase is considered to be a more ordered form of δ-MnO_2, containing a higher proportion of Mn_{II} ions (Buser *et al.* 1954). Buser (1959) provided a summary of this work and listed the Mn phases in nodules as 10 Å manganite,† 7 Å manganite and δ-MnO_2 by comparison with synthetic preparations. The phases differ in oxidation grade and in specific surface area (Buser & Graf 1955), δ-MnO_2 being significantly finer-grained and more highly oxidized than 7 Å manganite.

TABLE 2. MANGANESE OXIDE MINERALS IN FERROMANGANESE NODULES

(AFTER BURNS & BURNS 1977)

mineral	formula	crystal class	structure
todorokite	(Na, Ca, K, Ba, Mn^{2+}) $Mn_5O_{12} \cdot 3H_2O$	monoclinic	unknown, but may be related to hollandite and psilomelane
birnessite	$4MnO_2 \cdot Mn(OH)_2 \cdot 2H_2O$ or (Ca, Na)(Mn^{2+}, Mn^{4+})$_7O_{14} \cdot 3H_2O$	hexagonal	unknown, but may be related to chalcophanite
δ-MnO_2		hexagonal	possibly disordered, fine-grained birnessite

Subsequent work (see Manheim 1965) has shown that naturally occurring todorokite (Yoshimura 1934) is similar to 10 Å manganite and that birnessite (Jones & Milne 1956) has an identical X-ray diffraction pattern to that of 7 Å manganite. On the basis of the occurrence of these natural phases and their identification in ferromanganese nodules, Burns & Burns (1977) recommended the adoption of the nomenclatural scheme outlined in table 2. Note that Burns & Brown (1972) previously suggested that the X-ray reflexion at around 710 pm was a 101 plane of the 10 Å manganite rather than a discrete phase, and the term birnessite has been used for both birnessite (*sensu* Jones & Milne 1956) and δ-MnO_2 (Glemser *et al.* 1961; Bricker 1965; Cronan & Tooms 1969), and that todorokite is considered to be a mixture of *buserite* and its alteration products by Giovanoli *et al.* (1973) and Giovanoli & Bürki (1975).

The identification of the Mn phases in ferromanganese nodules by conventional X-ray diffraction methods presents some problems. Reflexions at around 245 and 140 pm are common to all three phases listed in table 2: it is not therefore possible to detect δ-MnO_2 in the presence of either of the other phases from line spacings alone. Burns & Burns (1977) recommend using peak intensities at 960, 710 and 245 pm to detect δ-MnO_2 in the presence of todorokite or birnessite. Lyle *et al.* (1977) have used this approach in a study of nodules from the southeastern Pacific. A separate problem surrounds the identification of birnessite, since a reflexion due to phillipsite, a most common accessory mineral in Pacific ferromanganese nodules, is also found at around 710 pm and the reflexion at 360 pm in birnessite is weak (Jones & Milne 1956). Although this phase is not as widely distributed in nodules as the other two phases, it does seem likely that previous identifications could be in error due to this interference. Figure 4 shows X-ray diffraction patterns of two nodule samples illustrating the problem.

† 1 Å = 10^{-10} m = 0.1 nm.

The regional distribution of the various Mn-bearing phases in nodules from the Pacific was first reported by Barnes (1967a). Glasby (1972a) has provided additional data, including that from Cronan (1967), and the combined information is shown in figure 5. It should be noted that only two phases are recognized here, namely todorokite and δ-MnO_2, because of the problems discussed above. Samples classified in the 7 Å manganite category by Barnes (1967b) are placed in the todorokite class, because they also contain todorokite and δ-MnO_2. It is apparent that todorokite occurs mainly in the eastern marginal Pacific and in two east-west zones in the northern and southern tropical areas. δ-MnO_2 occurs separately in the western Pacific. Although Barnes (1967a) also pointed out that δ-MnO_2 is the common phase in seamount nodules and suggested a hydrostatic pressure control on mineralogy, Glasby (1972b) has shown that the pressure range involved is too small to have an appreciable effect on the phases formed.

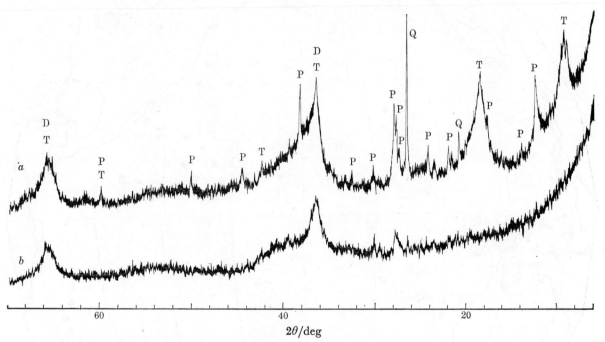

FIGURE 4. X-ray ($CuK\alpha$) powder diffraction patterns of ferromanganese nodules from the Pacific Ocean. (a) Sample WAH 18F3, 8° 16′ N, 152° 58′ W, water depth 5133 m. Contains todorokite (T) and δ-MnO_2 (D), as well as phillipsite (P) and quartz (Q). Chemical composition given in Calvert et al. (1978). (b) Sample AMPH 80, 11° 51′ S, 160° 51′ W, water depth 3803 m. Contains δ-MnO_2 and minor amounts of phillipsite and quartz. Chemical composition given in Calvert & Price (1977a).

Moreover, Skornyakova et al. (1975) and Calvert & Price (1977a) showed that some abyssal sea floor nodules in the central and southwestern Pacific, occurring at abyssal depths, contained only δ-MnO_2. It seems reasonable to conclude, therefore, that the mineralogy is governed by the environment of nodule formation rather than water depth *per se* (Cronan & Tooms 1969; Price & Calvert 1970; Glasby 1972a).

The mineralogy of the Mn phase in nodules is an important determinant of their minor element composition (Barnes 1967a; Cronan & Tooms 1969). Thus, nodules containing todorokite appear to contain, for example, more Cu, Mo, Ni and Zn, and nodules containing only δ-MnO_2 have more Ce, Co, Pb and Ti, as discussed in §2.2. Although some of this variation may be produced by the very different environments represented by the abyssal seafloor and seamounts, where Co enrichment is particularly marked, figure 6 shows that the mineralogy significantly influences the Ni content of abyssal sea floor nodules. These observations suggest, therefore, that we would expect some regional variation in the minor element composition of nodules and that this would be basically like that shown in figure 5. This has been confirmed by Cronan & Tooms (1969), Calvert & Price (1977a) and Piper & Williamson (1977) and will be discussed in §2.5.

FIGURE 5. Distribution of todorokite and δ-MnO$_2$ in nodules from the Pacific Ocean. Data from Barnes (1967b) and Glasby (1972a). Samples labelled ● contain todorokite and δ-MnO$_2$; samples labelled △ contain only δ-MnO$_2$. Samples included in a '7Å manganite' category by Barnes (1967a, b) are designated todorokite in this figure, since they contain both todorokite and δ-MnO$_2$ and because of the possible interference by phillipsite in the identification of this phase.

2.3.2. *Iron phases*

The mineralogy of the iron-bearing phase or phases in ferromanganese nodules is much less well known. Buser & Grütter (1956) identified goethite (α-FeOOH) in hydroxylamine-insoluble residues from oceanic nodules. Geothite has been reported most commonly in shallow-water nodules (Manheim 1965; Calvert & Price 1977b), whereas later reports on oceanic nodules have described the iron phase as amorphous (Goodell *et al.* 1971; Glasby 1972c). It is possible that the Fe is present as a hydrated ferric oxide polymer (Towe & Bradley 1967) which may be identical to a natural ferric gel (Coey & Readman 1973) and the mineral ferrihydrite, 2.5 Fe_2O_3 . 4.5 H_2O (Chukrov *et al.* 1973), possessing very short-range order and consequently being virtually amorphous to X-rays (Giovanoli & Bürki 1975; Burns & Burns 1977).

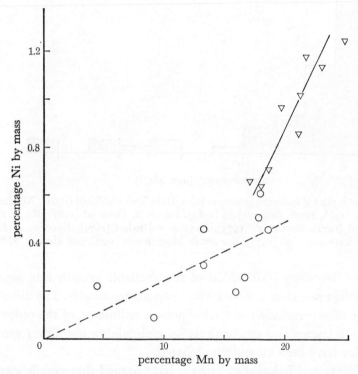

FIGURE 6. Relation between the Mn and Ni contents of ferromanganese nodules from the abyssal Pacific, illustrating the mineralogical control on minor metal concentrations (from Calvert & Price 1977a): $\triangledown$, todorokite; $\circ$, δ-MnO_2.

Burns & Burns (1975) have suggested that δ-MnO_2 and ferrihydrite (or Fe$_{III}$ oxide hydroxide) are probably isostructural and form epitaxial intergrowths in ferromanganese nodules. This association is considered to be responsible for the nucleation of oxyhydroxide precipitation and to inhibit the formation of more ordered structures in nodules. Burns & Burns (1975) have further argued that longer range ordering may proceed under high hydrostatic pressures to produce todorokite. This cannot be verified at present since it is well known that todorokite occurs in shallow-water nodules (see Calvert & Price 1977b) and that nodules containing only δ-MnO_2 and δ-MnO_2 together with todorokite occur at similar water depths in the equatorial Pacific (Calvert & Price 1977a). The important factor may be the abundance of iron in the nodules, which prevents ordering, since todorokite-free nodules have significantly lower Mn/Fe

ratios than todorokite-bearing nodules. A similar relation between the mineralogy and the
bulk composition of nodules has been noted by Lyle, Dymond & Heath (1977).

2.4. *Growth rates*

The rate of growth of ferromanganese nodules has been measured by a number of radiometric techniques, including $^{230}Th/^{232}Th$, excess ^{230}Th, K/Ar, ^{10}Be, gross α-activity, and by
hydration rind dating. Rates measured by ^{230}Th and ^{10}Be techniques (Bhat *et al.* 1973) and by
^{230}Th and hydration rind methods (Burnett & Morgenstein 1977) appear to be concordant.

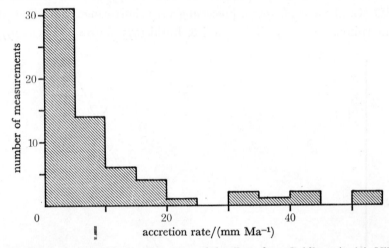

FIGURE 7. Measured growth rates of ferromanganese nodules. Data from Goldberg (1961), Nikolayev & Yefimova
(1963), Bender *et al.* (1966, 1970), Somayajulu (1967), Barnes & Dymond (1967), Ku & Broecker (1969),
Krishnaswami & Lal (1972), Bhat *et al.* (1973), Heye & Beiersdorf (1973), Kraemer & Schornick (1974)
Boulad *et al.* (1975), Sugimura *et al.* (1975), Burnett & Morgenstein (1976) and Heye & Marchig (1977).

Figure 7 shows the frequency distribution of the available growth rate measurement on
Pacific nodules. A median rate close to 5 mm Ma^{-1} appears reasonable. The difference between
this rate and the rate of accumulation of Pacific pelagic sediments, of the order of 2 m Ma^{-1}
(Ku *et al.* 1968), is well known and mechanisms for maintaining the slower growing nodules
at the sediment surface have been extensively debated.

Lalou & Brichet (1972) and Lalou *et al.* (1973 *b*) have argued that nodule growth rates are
in fact much higher than are derived from radiometric measurements. They maintain that
the ^{230}Th is deposited on the nodule surface after formation, the exponential decrease in activity
of the isotope in the near-surface layers of a few millimetres thickness being an artefact of this
process and not due to radioactive decay. Lalou *et al.* (1973 *b, c*) further argue that a higher
^{230}Th activity and the presence of ^{14}C in a phosphatized limestone nucleus in a Pacific nodule
is firm evidence for rapid growth, although this is contested by Bouland *et al.* (1975).

The suggestion that nodules grow episodically, periods of rapid growth alternating with
periods of dormancy, is also made by Lalou & Brichet (1972) and Lalou *et al.* (1973 *a*) and
explains some of the features of the microlaminated structure of nodules (Sorem & Foster
1972; Margolis & Glasby 1973). This is supported by some recent gross α-activity measurements
on Pacific nodules by Heye (1975) who has found that exceptionally Mn-rich zones have
growth rates in excess of 50 mm Ma^{-1} whereas Fe-rich zones represent periods of extremely
slow growth. Average accretion rates from 24 individual nodules were in the range 4–9 mm

Ma⁻¹ (figure 7). He also noted that buried nodules show no recent growth by the radiometric method used.

There appears to be a relation between nodule growth rates and their compositions (Piper & Williamson 1977; Heye & Marching 1977), Mn-rich nodules growing significantly more rapidly (figure 8). This may be due to the different modes of formation of nodules with different compositions, as discussed in §4.

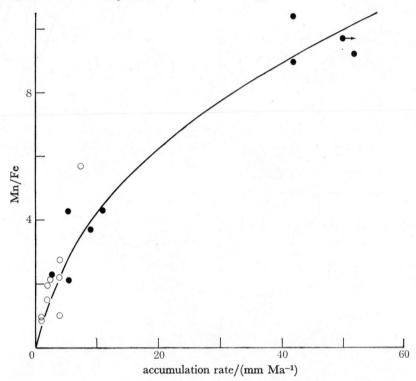

FIGURE 8. Relation between the composition, as expressed by the Mn/Fe ratio, and the growth rate of ferro-manganese nodules. Data from Krishnaswami & Lal (1972) (○) and Heye & Marchig (1977) (●).

2.5. *Regional variations in composition*

The chemical composition of ferromanganese nodules appears to vary from ocean to ocean (Mero 1965), and significant differences are found between nodules from different topographic or sedimentary provinces (§2.2). In addition, Mero (1962) discovered profound variations in the composition of nodules throughout the Pacific Ocean. Later work (Mero 1975; Cronan & Tooms 1969; Price & Calvert 1970; Calvert & Price 1977a; Piper & Williamson 1977) has confirmed this regional variation, which has entered into some of the recent discussions of the mode of formation of the nodules.

Mero (1962) pointed out that crude compositional regions could be recognized in the Pacific, different areas being characterized by Mn-rich, Cu- and Ni-rich, Fe-rich, and Co- and Pb-rich nodules. Price & Calvert (1970) considered a larger number of analyses of abyssal seafloor nodules and examined compositional differences throughout the Pacific by using element ratios, thereby removing the effects of diluent aluminosilicate material. They showed that the Mn/Fe ratio varied remarkably smoothly throughout the basin, high ratios occurring in the eastern marginal areas, and in an east–west belt centred at about 10° N, and low ratios occurring in the northwestern and southwestern areas. Figure 9 shows the distribution of this

FIGURE 9. Distribution of Mn/Fe ratios in ferromanganese nodules from the Pacific. Data from Willis (1970), Horn *et al.* (1973), Glasby *et al.* (1975), Meylan & Goodell (1976), Skornyakova (1976) and Calvert & Price (1977*a*).

ratio using more extensive data; the trends are very similar to those reported by Piper & Williamson (1977). Apart from the eastern marginal areas, high ratios are found in three east–west belts centred at approximately 10° N, 10° S and 40° S, while lowest ratios (below 1) are found in the northwestern and southwestern areas.

Regional variability is also evident in the concentration of some of the minor elements in ferromanganese nodules. Figure 10 shows that Ni is enriched in the abyssal areas where nodules are enriched in Mn and figure 11 shows that Co is enriched in the areas where the nodules have lower Mn contents relative to Fe. This is where the large seamount and atoll chains are found and reflects the enrichment of Co in nodules from this environment, as discussed in §2.2.

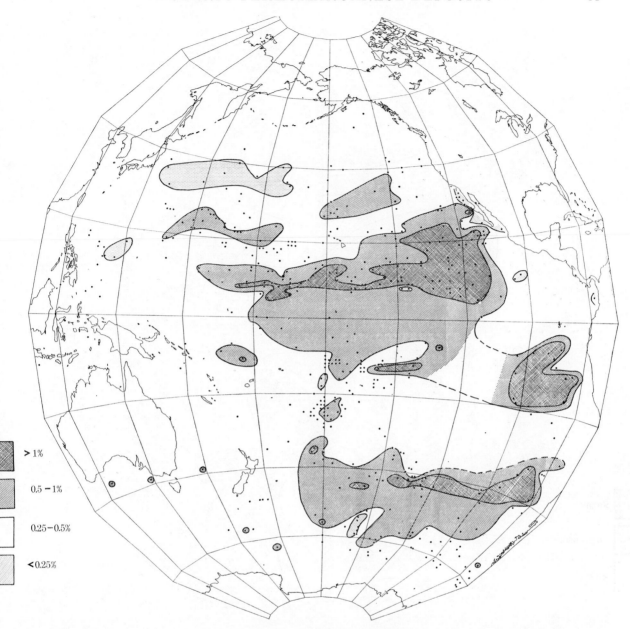

FIGURE 10. Distribution of nickel concentrations in ferromanganese nodules from the Pacific Ocean. Data sources as in figure 8. Copper shows a very similar distribution.

Using bulk chemical analyses, Calvert & Price (1977a) showed that the concentrations of a large group of major and minor elements, in addition to those shown in figures 10 and 11, also vary regionally in the Pacific since they are associated with either the Mn or the Fe phases of the nodules. Thus, Ba, Cu, Mo and Zn are related to the Mn contents and are highest in the northern equatorial region, whereas As, Ce, Pb, Sr, Y, Zr, Ti and P are related to the Fe contents and are highest where Mn/Fe ratios are lowest. These two groups of nodules also have different mineralogies (see figure 5). The Mn-rich nodules contain todorokite and δ-MnO$_2$, whereas the Fe-rich nodules contain only δ-MnO$_2$, the latter occurring on seamounts and on the abyssal sea floor.

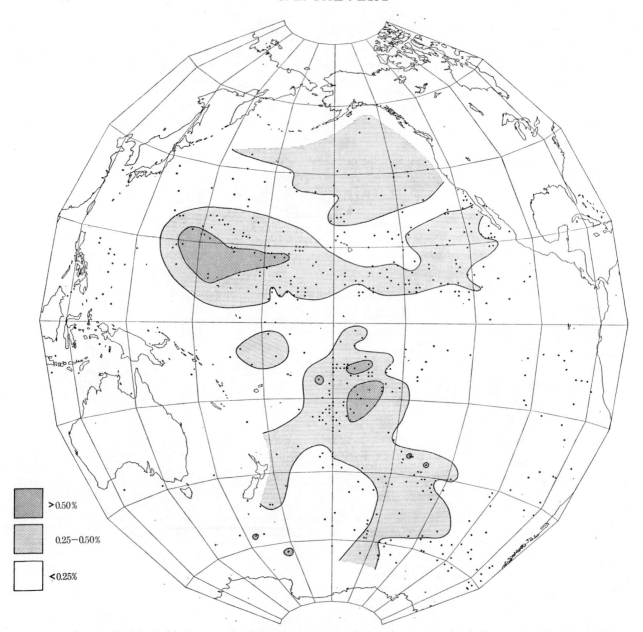

FIGURE 11. Distribution of cobalt concentrations in ferromanganese nodules from the Pacific Ocean. Data sources as in figure 8.

3. RIDGE CREST FERROMANGANESE DEPOSITS

3.1. *General*

Ferromanganese deposits of very variable form and composition are found sporadically on, or are associated with, the mid-ocean ridge system (figure 12). Revelle (1944) showed that the unconsolidated sediments in the southeastern Pacific contained high concentrations of Mn (see also El Wakeel & Riley 1961), and Skornyakova (1964) identified the East Pacific Rise (E.P.R.) area as a site of dispersed Mn and Fe enrichment. Subsequently, Boström & Peterson (1966) described the metal content of the E.P.R. sediments in some detail and further studies of these unusual sediments have been reported by Boström *et al.* (1969), Boström (1970) Boström & Peterson (1970), Bender *et al.* (1971), Piper (1973), Dymond *et al.* (1973), Dymond & Veeh (1975) and Heath & Dymond (1977).

Ferromanganese sediments also occur to the east of the E.P.R. in an area known as the Bauer Deep (B.D.) (Revelle 1944). They have been described by Dasch *et al.* (1971), Bischoff & Sayles (1972), Sayles & Bischoff (1973), Sayles *et al.* (1975), Bagin *et al.* (1975) and Heath & Dymond (1977).

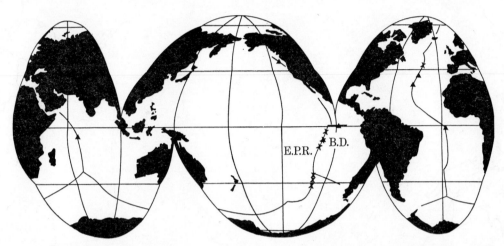

FIGURE 12. Occurrences of ferromanganese sediments, rocks and crusts on the mid-ocean ridge. E.P.R., East Pacific Rise; B.D., Bauer Deep; ▲, iron/manganese-rich crusts and crusts; ×, metalliferous sediment. Individual occurrences taken from Baturin (1971), Cronan (1972), M. R. Scott *et al.* (1974), Piper *et al.* (1975), Arcyana (1975), Bonatti *et al* (1976), Moore & Vogt (1976) and Burnett & Piper (1977).

As a result of deep drilling in the ocean basins (Joides 1967), ferromanganese-rich sediments are also known to occur above basaltic basement in many areas of the eastern Pacific (figure 13), as well as in the Atlantic and Indian Oceans. They are thought to have formed initially at active ridge crests and have since moved to their present positions as a result of sea-floor spreading. Their compositions have been reported by von der Borch & Rex (1970), von der Borch *et al.* (1971), Cook (1971), Drever (1971), Cronan *et al.* (1972), Cronan (1973, 1976*b*) and Cronan & Garrett (1973).

In addition to the unconsolidated ridge-crest sediments, ferromanganese crusts and rocks are also found on the mid-ocean ridge (figure 12). They represent isolated occurrences and have been recovered as a result of sporadic dredge sampling on topographic elevations or in fracture zones. More recently they have been systematically sampled from submersibles (Arcyana 1975).

3.2. *Chemical composition*

Representative analyses of ferromanganese deposits from ridge crest areas given in table 3 show that there is a much greater degree of variability in these deposits compared with ferromanganese nodules. The unconsolidated sediments have Mn/Fe ratios between 0.2 and 0.4 whereas the rocks and crusts have Mn/Fe ratios ranging from 0.002 to 39200. The aluminosilicate content of the deposits is quite low in the E.P.R. and basal metalliferous sediments and in the ironstones, whereas it is roughly the same as that in ferromanganese nodules in the B.D.

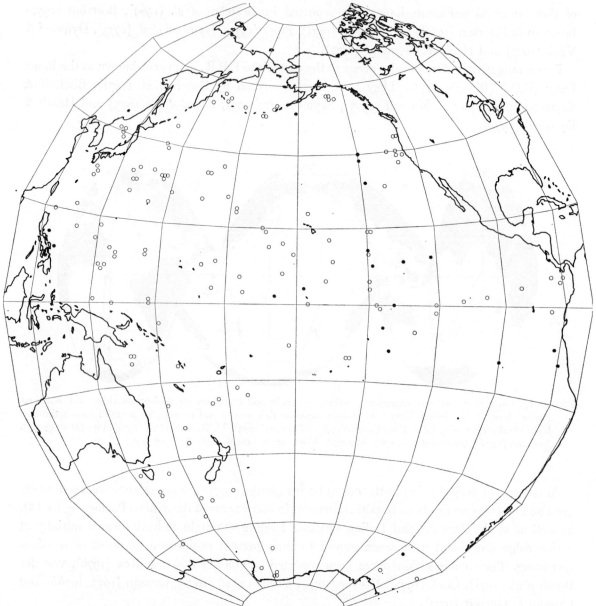

FIGURE 13. Distribution of basal metalliferous sediments in the Pacific Ocean. ○, D.S.D.P. drill sites;
●, metalliferous section present. Data taken from *Initial reports of the Deep Sea Drilling Project*.

sediments. Figure 14 shows that these deposits can be distinguished fairly easily from normal pelagic clays and from nodules.

Minor-element concentrations are also higher in the ridge-crest sediments compared with pelagic clays; in fact, As, Ba, Mo, Sr, Y, Zn and Zr contents are as high as those in nodules (table 3). Cobalt, Cu, Ni and Pb contents, on the other hand, are lower than those in nodules. Figure 15 shows that the deposits can also be distinguished from nodules on the basis of their Cu, Ni and Zn contents; nodules generally have Ni/Cu ratios greater than 1 whereas ridge crest sediments have Ni/Cu less than 1. Other elements enriched in ridge-crest sediments compared with pelagic clays include Ag, B, Cd, Hg, P, Tl and U (Boström & Peterson 1969; Boström & Fisher 1969; Fisher & Boström 1969; Horowitz 1970; Cronan 1972; Berner 1973).

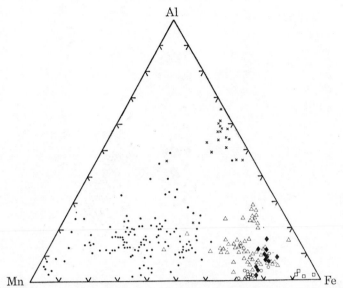

FIGURE 14. Composition of oceanic ferromanganese deposits on the basis of Mn–Fe–Al contents. Data sources: nodules (●), Mero (1965); E.P.R. (○) and B.D. (△) sediments, Heath & Dymond (1977); D.S.D.P. basal sediments (◆), S. E. Calvert, P. K. Studdart & K. Hampton (unpublished); ironstones (□), Bonatti & Joensuu (1966) and Piper *et al.* (1976); pelagic clays (×), Calvert & Price (1977a).

TABLE 3. CHEMICAL COMPOSITION OF METALLIFEROUS RIDGE-CREST SEDIMENTS,

CRUSTS AND ROCKS

(Major elements in percentages by mass; minor elements in parts/10^6.)

element	1	2	3	4	5	6	7	8	9	10
Si	6.12	17.43	19.5	8.50	—	6.8	8.6	—	—	5.6
Al	0.51	3.24	2.61	2.75	—	0.5	1.0	—	—	0.8
Ti	—	—	—	0.25	—	—	—	—	—	0.07
Fe	30.20	15.83	13.6	23.14	0.06	30.8	28.5	0.22	21.8	39.0
Ca	—	—	1.74	1.67	—	1.8	1.3	—	—	0.14
Mg	—	—	2.39	1.52	—	0.6	0.7	—	—	0.24
K	—	—	0.85	1.25	—	0.3	0.3	—	—	0.25
Mn	9.92	5.74	3.83	4.77	39.0	1.65	2.05	54.6	16.7	0.08
P	—	—	—	0.60	—	—	—	—	—	—
As	—	—	—	250	—	—	—	—	—	—
Ba	6000	18600	18266	1633	—	105	267	—	—	80
Co	—	—	176	113	19	62	5	34	508	28
Cu	1450	1171	1084	1043	43	85	6	51	735	35
Mo	—	—	—	233	—	—	—	—	—	—
Ni	642	1066	1021	721	353	317	54	181	1950	170
Pb	—	—	—	112	—	—	—	—	—	—
Rb	—	—	—	53	—	—	—	—	—	—
Sr	—	—	1178	873	—	583	—	—	—	—
Y	—	—	—	120	—	—	—	—	—	< 5
Zn	594	413	376	667	—	—	450	2023	543	—
Zr	—	—	—	133	—	—	—	—	—	60

Sources: 1, East Pacific Rise sediment, mean of 7 samples on a carbonate-free basis (Heath & Dymond 1977, Table 2). 2, Bauer Deep sediment, mean of 7 samples on a carbonate-free basis (Heath & Dymond 1977, Table 2). 3, Bauer Deep sediment, mean of 56 samples on a carbonate-free basis (Sayles *et al.* 1975, Table 3). 4, Basal metalliferous sediments, mean of 9 samples, on a total sample basis, from D.S.D.P. Sites 37, 39 and 66, eastern Pacific (S. E. Calvert, P. K. Studdart & K. Hampton, unpublished data). 5, Manganese crust from Tag hydrothermal field 27° N Mid-Atlantic Ridge; mean of 7 samples (M. R. Scott *et al.* 1974, Table 2). 6, Ironstone from Station AMPH D2, 10° 38′ S, 109° 36′ W, 1790–2130 m depth, East Pacific Rise; mean of 3 samples (Bonatti & Joensuu 1966, Table 1). 7, Ironstone from Dellwood Seamount, Northeast Pacific; single sample (Piper *et al.* 1975, Table 1). 8, Manganese crust from Galapagos Spreading centre, 02° 42′ N, 95° 13′ W, 2562 m depth; mean of 7 samples (Moore & Vogt 1976, Table 1). 9, Ferromanganese crust from Galapagos Spreading Centre, 02° 17.8′ N, 101° 1.5′ W, 3150 m depth; single sample (Burnett & Piper 1977, Table 1). 10, Sulphide concretion from Romanche Fracture Zone, Mid-Atlantic Ridge; single sample (Bonatti *et al.* 1976, Table 1).

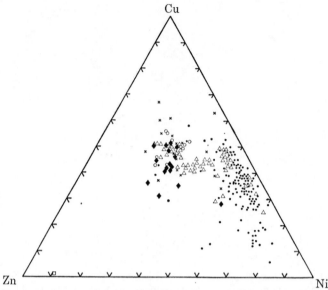

FIGURE 15. Composition of oceanic ferromanganese deposits on the basis of Cu–Ni–Zn contents. Data sources as in Fig. 13.

In contrast to the metal-enrichment in ridge-crest sediments, the rocks and crusts are markedly deficient in minor metals (table 3). The available data also suggest that Ni/Cu ratios are greater than unity.

The partition of some of the major and minor elements between different mineral phases (see §3.3) in the ridge-crest and basal metalliferous sediments has been examined using leaching procedures by Cronan & Garrett (1973), Sayles & Bischoff (1973), Sayles et al. (1975), Cronan (1976b) and Heath & Dymond (1977). In the E.P.R. sediments, Mn, Fe, Co, Ni, Cu and Zn are mainly present in oxyhydroxide phases whereas in the B.D. significantly more Fe is located in a smectite phase. In the basal metalliferous sediments, Cu and Zn appear to be mainly associated with goethite whereas a higher proportion of the Pb and almost all the Co and Ni are associated with Mn oxyhydroxides. Eklund (1974) has examined the composition of discrete todorokite, smectite and mixed oxide/silicate particles in B.D. sediments and has confirmed these patterns.

The distributions of the lanthanides in ridge-crest ferromanganese deposits show some distinct differences from those in ferromanganese nodules (§2.2). In the unconsolidated sediments, the absolute concentrations of the lanthanides are lower than in pelagic clays and the shale-normalized pattern is very similar to that of seawater (figure 16). This has been used as firm evidence for the derivation of the lanthanides from seawater without significant fractionation. However, Dymond et al. (1973) have suggested that the presence of significant amounts of phosphatic fish debris, which concentrates lanthanides from seawater (Arrhenius & Bonatti 1965), could affect the lanthanide patterns in such sediments, although Piper & Graef (1974) could not corroborate this. Eklund (1974) has shown that fish debris in Bauer Deep sediments contains less than 100 parts $Ce/10^6$ and 2400–3500 parts $La/10^6$ and, on the basis of the abundance of this material in these sediments, has suggested that the low Ce/La ratio in the bulk sediment may simply reflect mixing of components with different lanthanide contents.

The lanthanide patterns of ridge-crest ironstones and crusts appears to be highly variable (figure 16). A pattern similar to that of seawater has been obtained for the Mn-rich crust from the Galapagos Spreading Centre (Burnett & Piper 1977), whereas a pattern similar to that of basalt is obtained for the Dellwood Seamount ironstone (Piper *et al.* 1975). In the case of the Romanche Trench sulphide concretion, the pattern is quite unlike either of the other distributions.

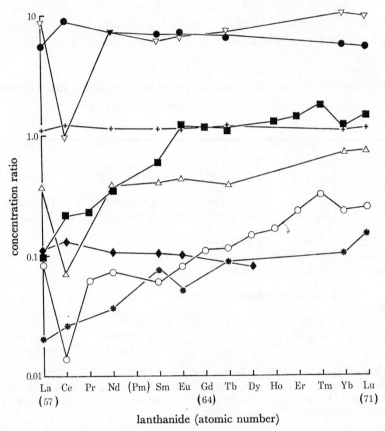

FIGURE 16. Distribution of the lanthanides in ridge-crest ferromanganese deposits (from Piper & Graef 1974; Piper *et al.* 1975; Bonatti *et al.* 1976; Burnett & Piper 1977) nodules and pelagic clay (from Piper 1974*a*), seawater (from Høgdahl *et al.* 1968) and Mid-Atlantic Ridge (M.A.R.) basalt (from Frey & Haskin 1964): *, Dellwood Seamount ironstone; ●, ferromanganese nodules; ▽, Galapagos Spreading Centre crust; +, pelagic clay; △, E.P.R. sediment; ■, M.A.R. basalt; ◆, Romanche Trench sulphide; ○, seawater (10^6 × concentrations). Concentration ratio is the ratio of each individual lanthanide to its concentration in average shale (Piper 1974*a*).

The concentration and isotopic composition of uranium in ridge-crest sediments, rocks and crusts also provide information on the likely sources of elements in such deposits (table 4). Absolute concentrations of U in the sediments are significantly higher than those in pelagic clays (Heye 1969; Fisher & Boström 1969; Bender *et al.* 1971; Veeh & Boström 1971), and the $^{234}U/^{238}U$ ratio in surface sediments is close to that in seawater, namely 1.15 (Thurber 1962). This probably indicates that the U was derived from seawater by coprecipitation onto Mn and Fe oxyhydroxides (Bender *et al.* 1971; Veeh & Boström 1971).

In ridge-crest crusts and ironstones, the distribution of U appears to fall into two distinct groups. In Mn-rich crusts from the TAG field (M. R. Scott *et al.* 1974), U concentrations are quite high and the $^{234}U/^{238}U$ ratio is not significantly different from that of seawater, and a similar conclusion has been reached for the Mn crust from the Galapagos Spreading Centre

(Burnett & Piper 1977). In distinct contrast, ridge-crest ironstones have significantly less U than pelagic clays and anomalously high $^{234}U/^{238}U$ ratios (Veeh & Boström 1971; Piper *et al.* 1975). It appears that whereas the U in the Mn-rich crusts is derived from seawater, that in the ironstones has suffered some fractionation, possibly by the leaching of ^{234}U from basaltic rocks in a hydrothermal system (Rydell & Bonatti 1973). Kazachevskii *et al.* (1964) reported $^{234}U/^{238}U$ ratios of 1.2 in limonitic precipitates from the Kurile Islands and 1.14 in hydrous Mn and Fe oxide suspensions from the Banu-Wuhu volcano in Indonesia (see Zelenov 1964).

TABLE 4. CONCENTRATIONS OF URANIUM AND ISOTOPIC COMPOSITIONS OF URANIUM, STRONTIUM AND LEAD IN RIDGE-CREST FERROMANGANESE DEPOSITS

deposit	U parts/10⁶	$^{234}U/^{238}U$	$^{87}Sr/^{86}Sr$	$^{206}Pb/^{204}Pb$	$^{207}Pb/^{204}Pb$
E.P.R. sediments	4.16–11.7[1]	1.08–1.14[1]	—	—	—
E.P.R. sediments	4.55–6.90[2]	1.03–1.15[2]	0.708 ± 0.001[2]	18.5_x[2,4]	15.543[2]
E.P.R. sediments	0.29–0.71[3]	1.06–1.16[3]	—	—	—
B.D. sediments	1.45–2.14[3]	1.00–1.04[3]	0.70899[10]	18.306[10]	15.518[10]
basal metalliferous sediments	1.97–5.83[4]	0.99–1.04[4]	0.7078–0.7099 ± 0.0001[4]	18.395[4]	15.561[4]
ironstone, AMPH D2	1.25[1]	1.21–1.22[1]	—	—	—
ironstone, AMPH D2	1.25–2.39[5]	1.15–1.23[5]	—	—	—
ironstone, Dellwood Seamount	0.25[6]	1.29[6]	—	—	—
manganese crust Tag area, M.A.R.	9.25–16.46[7]	1.06–1.23[7]	—	—	—
pelagic clays	1–3[8]	0.92[8]	0.72[11]	—	—
seawater	0.003[9]	1.15[9]	0.70906[12]	18.812[14]	15.63[14]
E.P.R. basalt	—	—	0.70248[13]	18.31[15]	15.55[15]

Sources of data: 1, Veeh & Boström (1971); 2, Bender *et al.* (1971); 3, Dymond & Veeh (1975); 4, Dymond *et al.* (1973); 5, Rydell & Bonatti (1973); 6, Piper *et al.* (1977); 7, M. R. Scott *et al.* (1974); 8, Heye (1969); 9, Thurber (1962); 10, Dasch *et al.* (1971); 11, Dasch (1969); 12, Hildreth & Henderson (1971); 13, Hart (1971); 14, Reynolds & Dasch (1971); 15, Tatsumoto (1966).

The isotopic composition of strontium in ridge crest sediments (table 4) has a value similar to that of seawater (Bender *et al.* 1971; Dasch *et al.* 1971). This has also been taken to imply a seawater source for this element, although Dymond *et al.* (1973) have found evidence that exchange of strontium between solid and aqueous phases in 30–90 Ma old basal metalliferous sediments has probably altered the original isotopic composition, thereby obscuring its source.

Available lead isotopic compositions of ridge-crest sediments (table 4) fall in the range exhibited by ocean ridge basalts (Tatsumoto 1966) and are distinctly different from those of ferromanganese nodules, assumed to reflect the seawater ratios. This evidence is thought to indicate an ultimate magmatic or hydrothermal origin for the metalliferous sediments (Bender *et al.* 1971; Dasch *et al.* 1971; Dymond *et al.* 1973), although the data do not allow a distinction to be made between direct precipitation from hydrothermal solutions and other types of solution containing Pb from a magmatic source.

3.3. *Mineralogy*

The mineralogical composition of the oxyhydroxide component of ridge-crest deposits appears to be more variable than that of the nodules. Although the modern E.P.R. sediments were initially described as amorphous (Boström & Peterson 1966), there is some indication from the work of Bagin *et al.* (1975) that goethite and an Fe-rich smectite are both present. In the case of the B.D. sediments, more extensive investigations by Dasch *et al.* (1971), Sayles & Bischoff (1973), Dymond *et al.* (1973), Eklund (1974), Sayles *et al.* (1975) and McMurtry (1975) have established the presence of δ-MnO_2 and todorokite (§2.3.1) in addition to fine-grained goethite (figure 17), and an Fe-rich smectite, probably nontronite.

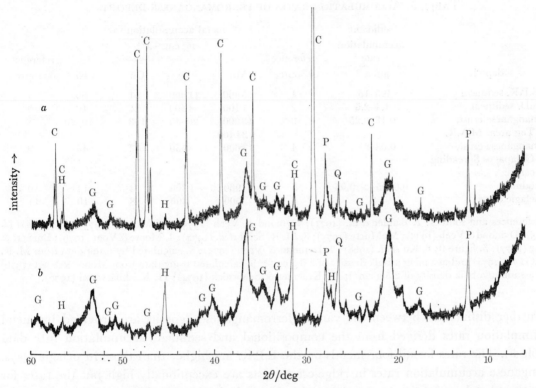

FIGURE 17. X-ray powder diffraction patterns of (*a*) Bauer Deep sediment and (*b*) basal metalliferous sediment (D.S.D.P. site 39). Patterns obtained by ultra-slow scanning, using CuKα radiation and a graphite monochromator. P, Phillipsite; G, goethite; C, calcite; Q, quartz; H, halite. Sample *a* kindly provided by J. Dymond, Oregon State University; sample *b* kindly provided by the Deep Sea Drilling Project, National Science Foundation, Washington, D.C.

Basal metalliferous sediments from the eastern Pacific have also been described as amorphous (von der Borch *et al.* 1971) although subsequent studies have shown the presence of goethite (Dymond *et al.* 1973; Drever 1971) and occasional psilomelane and Fe-rich smectite (Dymond *et al.* 1973). Figure 17 shows the X-ray evidence for the presence of crystalline goethite in such sediments.

Of the two types of ridge-crest rocks and crusts, the Mn-rich variety is reported to contain δ-MnO_2 (Burnett & Piper 1977), birnessite and todorokite (Scott *et al.* 1974; Moore & Vogt 1976) while the ironstones consist of goethite (Bonatti & Joensuu 1966) or X-ray amorphous iron oxyhydroxides (Piper *et al.* 1975).

The main mineralogical contrast between ferromanganese nodules and ridge crest deposits appears to lie in the common occurrence of goethite and the highly variable occurrences of readily identifiable Mn phases in the latter.

3.4. *Accumulation rates*

The total accumulation rate data in table 5 indicate that the unconsolidated E.P.R. sediments accumulate more rapidly than the B.D. sediments and pelagic clays. The Mn-rich crusts accumulate somewhat more slowly but significantly more rapidly than ferromanganese nodules.

TABLE 5. ACCUMULATION RATES OF FERROMANGANESE DEPOSITS

deposit	sediment accumulation rate μm a^{-1}	reference or source	Mn	Fe	Cu	Ni	reference or source
			metal accumulation rate ng cm^{-2} a^{-1}				
E.P.R. sediment	9.3–15	1	5800	11000	51	56	7
B.D. sediment	1.4–2.5	2	1100	4400	25	16	7
manganese crust, Tag area, M.A.R.	0.13–0.25	3	13000–24400	20–40	1–3	10–20	8
manganese crust, Galapagos Spreading centre	0.08–2	4	13600	550	13	45	9
nodules	0.0005–> 0.05	5	960	1300	4	14	10
pelagic clays	2	6	500	800	8	10	11

References and sources: 1, Bender *et al.* (1971); Dymond & Veeh (1975); McMurtry (1975); 2, Sayles *et al.* (1975); Dymond & Veeh (1975); McMurty (1975); 3, M. R. Scott *et al.* (1974); 4, Moore & Vogt (1976); Burnett & Piper (1977); 5, figure 9; 6, Ku *et al.* (1968); 7, Dymond & Veeh (1975); 8, calculated by using data from M. R. Scott *et al.* (1974) and assuming a bulk density of 2.5 g cm^{-3}; 9, calculated using data from Moore & Vogt (1976) and assuming a bulk density of 2.5 g cm^{-3}; 10, Kraemer & Schornick (1974); 11, Krishnaswami (1976).

Further distinctions between the various ferromanganese deposits are provided by metal accumulation rates derived from the compositional and sediment accumulation rate data (table 5). The rates for all four metals on the E.P.R. are higher than those in pelagic clays. Manganese accumulation rates in ridge-crest crusts are exceptionally high but the rates for the other metals are of the same order as those for pelagic clays. The close similarity between the accumulation rates of Mn in nodules and sediments has been discussed by Bender *et al.* (1970). Accumulation rates of ridge-crest ironstones are not available.

4. FORMATION OF OCEANIC FERROMANGANESE DEPOSITS

An explanation of the formation of the ferromanganese deposits of the deep sea involves two separate problems. The ultimate sources of the metals in the deposits are of fundamental interest, whereas the modes of accretion of the oxyhydroxide phases on the sea floor, regardless of sources, are of immediate concern. The sources of the metals are considered to be continental denudation and marine volcanism and the relative importance of these two sources had led to considerable controversy since they were first discussed by Murray & Renard (1891). It seems clear that both sources are important (Cronan 1976), since mass balance estimates (Horn & Adams 1966; Boström 1967; Varentsov 1971; Elderfield 1976) indicate that a significant fraction of the Mn in oceanic deposits has a continental source and because recent studies of hydrothermal systems (Ellis 1973) and the reaction between fresh basaltic rock and seawater (Corliss 1971; Hajash 1975; Bischoff & Dickson 1975) show that several metals of concern can be derived from within the ocean basins.

The mode of formation of ferromanganese nodules involes the slow precipitation of oxyhydroxides on exposed solid surfaces, some of which may become nuclei for the formation of concretions. Goldberg & Arrhenius (1958) suggested that a ferric oxide surface could provide an initial reaction site on which seawater Mn$_{II}$ could be oxidized by molecular oxygen. Stumm & Morgan (1970) have shown that the precipitation of ferric hydroxide from iron hydroxide species, the forms of importance in seawater (Byrne & Kester 1976), is much more rapid than the precipitation of Mn$_{IV}$ oxides at the same pH. The initial deposition of an Fe oxyhydroxide film on foraminiferal and coral surfaces in nodules has been observed by Burns & Brown (1972). The autocatalytic precipitation of Mn oxyhydroxides (Morgan & Stumm 1970; Hem 1976) on such surfaces would then lead to continual accretion of mixed oxyhydroxides from seawater, the rate of deposition being controlled by the rate at which the reactants are brought to the surfaces.

Burns & Brown (1972) suggested that 10 Å manganite (todorokite) may be the first formed Mn-phase on the sea floor which could yield a separate, more oxidized phase (δ-MnO$_2$), accompanied by dehydration and shrinkage. On the other hand, Burns & Burns (1975) suggested that the epitaxial intergrowth of δ-MnO$_2$ and Fe$_{III}$ oxyhydroxide is the key to the nucleation of nodule growth and that todorokite forms subsequently under high hydrostatic pressure. This seems unlikely for the reasons given in §2.3.2.

Some additional insight into the possible modes of formation of oceanic nodules comes from recent work on the chemical and mineralogical compositions of nodules throughout the Pacific Ocean (§2.5) and on the modes of formation of nearshore nodules (Calvert & Price 1977b). The regional geochemical variation in the Pacific (figures 9–11) led Price & Calvert (1970) to suggest that two distinct mechanisms of formation are involved, a precipitation from normal seawater, which produces a deposit consisting of δ-MnO$_2$ with roughly equal Mn and Fe and relatively high Ce, Co, Pb and Ti contents, and a diagenetic precipitation producing a deposit consisting of todorokite with more Mn than Fe and high Cu, Ni, Mo and Zn contents. Nearshore nodules, consisting of todorokite, are known to form by the precipitation of remobilized metals from expelled sediment pore waters which has the effect of producing extreme Mn and Fe fractionations (Manheim 1965; Cheney & Vredenburg 1968; Calvert & Price 1970). The explanation of Price & Calvert (1970) was an attempt to extend this mechanism to the abyssal nodules, although it is recognized that anoxic conditions, which promote transition metal mobility in rapidly accumulating nearshore sediments, do not exist in pelagic clays.

Evidence for two distinct modes of precipitation of metals in abyssal nodules is provided, however, by the important observation of Raab (1972) that discoidal nodules from the abyssal northern equatorial Pacific had compositionally distinct upper and lower surfaces. Upper surfaces were enriched in Fe, Co and Pb, while lower surfaces contained more Mn, Ni, Cu, Mo and Zn. Ku & Broecker (1969) and Lalou & Brichet (1972) have also pointed out that upper and lower surfaces of nodules often have distinctly different α-activities and Kunzendorf & Friedrich (1976) observed higher concentrations of U and Th on the surfaces of nodules last exposed to seawater.

Calvert & Price (1977a) used the observations of Raab (1972) to suggest that the upper surfaces of the nodules represent precipitation predominantly from seawater whereas the lower surfaces are diagenetic. They further proposed, on the basis of bulk compositions, that there is a continuous variation in the relative proportions of these two sources on a regional scale; the seawater end-member is perhaps best represented by oceanic seamount nodules, which grow slowly on rock or coral substrates, and the diagenetic end-member is represented by the Mn-enriched nodules in the northern equatorial Pacific, which grow significantly more rapidly (figure 8) on a reactive, oxidized sediment substrate. Because of the existence of diagenetic as well as normal seawater metal sources, ultimate sources become obscured.

The distribution of the diagenetic abyssal nodules in the Pacific (figure 9) is clearly controlled by processes taking place in the underlying sediments. Although Price & Calvert (1970) maintained that these nodules form on more rapidly accumulating sediments where diagenetic reaction would more marked, Strakhov (1974a) and Piper & Williamson (1977) have shown

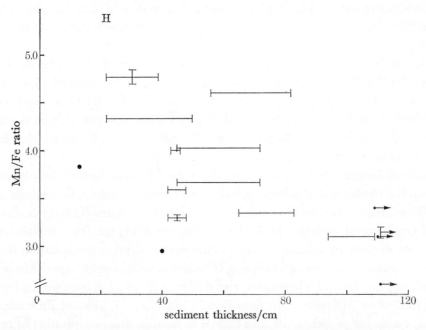

FIGURE 18. Relation between the Mn/Fe ratio of the nodules and the thickness of Quaternary sediment in a small survey area in the northern equatorial Pacific. Horizontal bars indicate the uncertainty in the last appearance of Quaternary radiolarians in the sediments. Vertical bars indicate the range of duplicate analyses of single nodules. Arrows indicate the absence of pre-Quaternary sediment in a core. After Calvert et al. (1978).

that the sediments probably accumulate very slowly in these areas, the fractionation of the metals most probably taking place at or close to the sediment surface. This inferred relation between the composition of the nodules and the accumulation rate of the associated sediment has been corroborated by Calvert et al. (1978) in a suite of Mn-enriched nodules from a small survey area of the northern equatorial Pacific (figure 18). The mechanism of fractionation of Mn and Fe at the sediment surface is not understood but may involve the reaction of Fe oxyhydroxides with biogenous silica to form a smectite phase in the sediment, as suggested by Lyle et al. (1977) who have further proposed that the removal of Fe from the mixed nodular precipitate would facilitate the crystallization of todorokite, the phase in which divalent minor metals are readily accommodated.

The relation between the composition of abyssal ferromanganese nodules and their growth rates (§2.4) may provide an alternative explanation for the formation of Mn-enriched nodules in the deep sea. It appears from the work of Heye (1975) that Mn-rich zones in nodules from the northern equatorial Pacific accumulate more rapidly than zones with higher Fe contents.

Hence, nodules with high bulk Mn/Fe ratios may simply represent precipitates which have received additional, relatively rapid impulses of Mn superimposed upon the 'normal' accretion of the mixed Mn and Fe oxyhydroxides from seawater.

The enrichment of Ni and Cu in the Mn-rich nodules from the northern equatorial Pacific (figures 9 and 10) has been explained by their additional supply by settling biological debris (Greenslate et al. 1973). This material, mainly skeletal, containing trace metals scavenged from the water column, dissolves on the sediment surface below the lysocline (Adelseck & Berger 1975) thereby releasing a suite of adsorbed and occluded metals (see Piper & Williamson 1977). Determinations of the fluxes of metals to the sea floor by such a process are required to test this hypothesis, but the locations of the zones of transition-metal enriched nodules (figure 9) in areas of high primary biological production in the surface waters of the Pacific (Reid 1962; Koblentz-Mishke et al. 1970; El Sayed 1970) are certainly consistent with such an explanation.

The rôle of microorganisms in the precipitation of Mn and Fe on the sea floor has been discussed for a considerable time. Ehrlich (1966) has isolated bacteria from nodules and has suggested that they promote the oxidation of MnII to MnIV (see also Sorokin 1972). Although bacteria are involved in the precipitation of Mn and Fe oxyhydroxides in lakes (Perfil'ev et al. 1965) their participation in oxide precipitation in the sea has not been demonstrated. Stumm (in Ehrlich 1964) and Sorokin (1972) have cautioned against the acceptance of biological agencies when compounds that are so readily oxidized inorganically are involved.

Although a direct biological involvement in the formation of ferromanganese nodules remains conjectural, it is clear that microorganisms can play an important part in constructing ferromanganese concretions. Wendt (1974) and Greenslate (1974a, b) have shown that agglutinating Foraminifera, probably Saccorhiza, and other shelter-building organisms, are abundant on the surfaces and in the interiors of some nodules. They have suggested that the skeletal structures of the organisms contribute to the overall growth of nodules and provide a framework in which metal oxyhydroxides can be deposited. A variation on this theme is the suggestion of Monty (1973) that the nodules represent bacterial stromatoliths.

The formation of ferromanganese deposits on the ocean ridge-crest is probably more directly linked to hydrothermal activity at constructive plate margins (Bonatti 1975). The principal features of such deposits, suggesting such an influence, include their rapid accumulation rates and their extreme compositional variability. Moreover, recent work on the thermal régimes at such plate margins (Lister 1972; Spooner & Fyfe 1973; Williams & von Herzen 1974), together with observational (Corliss 1971; Hart 1973) and experimental (Hajash 1975; Bischoff & Dickson 1975; Elderfield et al. 1977) studies of the alteration of basaltic rocks by seawater, have suggested that extensive alteration of freshly intruded rock takes place to considerable depths and that the fluxes of metals into the circulating seawater are important in the marine geochemical balance (Wolery & Sleep 1976; see also Strakhov 1974b; Elderfield 1976; Lyle 1976). There is at present no report of any direct observation of hydrothermally altered seawater emanating at ocean ridges, although positive temperature anomalies have been recorded in the Tag area of the Mid-Atlantic Ridge (Rona et al. 1975) and on the Galapagos Spreading Centre (Detrick et al. 1974; Weiss et al. 1977). Lupton et al. (1977) have recently discovered excess ^{3}He and ^{222}Rn anomalies in water samples collected in high-temperature plumes on the Galapagos Spreading Centre, indicating the hydrothermal origin of the plumes, and Klinkhammer et al. (1977) have demonstrated the presence of high concentrations of Mn in these same plumes and in the water over the ridge crest.

In some geothermal areas, such as at Wairakei in New Zealand (Ellis 1970) and in the Reykjanes system in Iceland (Björnsson *et al.* 1972), many features produced by rock/hot water interaction may also be applicable to the mid-ocean ridge setting (Wolery & Sleep 1976). In particular, the circulating fluid, which in Iceland is seawater, contains high concentrations of Mn and Fe. The results of experiments on the hydrothermal alteration of basaltic material by seawater demonstrate that Mn, Fe, Cu and Ni, as well as a wide range of major elements, are released in significant quantities to the aqueous phase (Mottl *et al.* 1974; Hajash 1975; Bischoff & Dickson 1975). On the basis of mass balance estimates, it is thought that such a transfer could account in major part for the mass of metalliferous sediment on the ridge system.

The precipitates from the hydrothermal system are evidently fractionated to a considerable degree. Iron is much more abundant in the unconsolidated sediments and occurs as virtually pure ironstones in some cases. The lanthanide patterns and the U isotopic compositions of these ironstones point to a hydrothermal source. On the other hand, the lanthanides, U and Sr probably have a seawater source in the unconsolidated sediments whereas Pb appears to be magmatic. The sources of the other constituents of the deposits are deduced from their accumulation rates. Thus, Mn is accumulating more rapidly than in pelagic clays in the E.P.R. sediments and in the isolated Mn-rich crusts, although a significant fraction of this element could be derived from seawater in the B.D. For the minor transition metals, Dymond & Veeh (1975) contended that both hydrothermal and seawater sources could be involved; and Heath & Dymond (1977) concluded, on the basis of their partition patterns, that whereas more than 50 % of the Ba, Cu, Ni and Zn are hydrothermal on the E.P.R. significantly more of the Ba and Ni have a seawater source in the B.D.

Discussion of the likely source of P in the E.P.R. sediments shows that evidence for element sources based on accumulation rates can be equivocal. Berner (1973) showed that P is enriched in E.P.R. sediments, deduced that it was also accumulating rapidly at this site and that its source was seawater. Froelich *et al.* (1977) have confirmed these observations by showing that the P is present in the dispersed oxyhydroxide phase and is accumulating approximately 40 times faster on the rise crest than in the B.D. where the rate can be shown to be close to that in pelagic clays. In spite of this high relative rate, which incidentally is even higher than the relative rates for the metals (table 5), Froelich *et al.* (1977) maintain that the P is most probably derived from seawater. The key process involved would appear to be the high sorptive capacity of the poorly ordered oxyhydroxides, which may be derived from hydrothermal solutions, but which scavenge other constituents at a very high rate from circulating seawater. It seems possible that the concentrations of minor transition metals observed in the ridge-crest sediments could also be due to their uptake at high rates from normal seawater by such reactive components. Analyses of the metal contents of hydrothermal solutions at the ridge-crest and some less equivocal methods for identifying ultimate metal sources in the sediments are clearly necessary to resolve this interesting problem.

REFERENCES (Calvert)

Adelseck, C. G. & Berger, W. H. 1975 *Cushman Found. Foram. Res., spec. Publ.* **13**, 70–81.
Ahrens, L. H., Willis, J. P. & Oosthuizen, C. O. 1967 *Geochim. cosmochim. Acta* **31**, 2169–2180.
Arcyana 1975 *Science, N.Y.* **190**, 108–116.
Arrhenius, G. & Bonatti E. 1965 *Progr. Oceanogr.* **3**, 7–21.
Bagin, V. I., Bagina, O. A., Bogdanov, Y. A., Gendler, T. S., Lebedev, A. I., Lisitsin, A. P. & Pecherskiy, D. M. 1975 *Geochim. Int.* **12**, 105–125.

Barnes, S. S. 1967a *Science, N.Y.* **157**, 63–65.

Barnes, S. S. 1967b Ph.D. thesis, University of California, San Diego, 59 pages.

Barnes, S. S. & Dymond, J. R. 1967 *Nature, Lond.* **213**, 1218–1219.

Baturin, G. N. 1971 In *The history of the world ocean* (ed. L. A. Zenkevitch), pp. 259–277. Moscow: Nauka.

Bender, M., Broecker, W. S., Gornitz, V., Middel, U., Kay, R. S., Sun, S. S. & Biscaye, P. 1971 *Earth planet. Sci. Lett.* **12**, 425–433.

Bender, M. L., Ku, T.-L. & Broecker, W. S. 1966 *Science, N.Y.* **151**, 325–328.

Bender, M. L., Ku, T.-L. & Broecker, W. S. 1970 *Earth planet. Sci. Lett.* **8**, 143–148.

Berger, W. H. 1970 *Mar. Geol.* **8**, 111–138.

Berner, R. A. 1973 *Earth planet. Sci. Lett.* **18**, 77–86.

Bhat, S. G., Krishnaswamy, S., Lal, D., Rama & Somayajulu, B. L. K. 1973 In *Proc. Symp. on Hydrogeochemistry and Biogeochemistry* (ed. E. Ingerson), vol. 1, pp. 443–462. Washington, D.C.: The Clarke Co.

Bischoff, J. L. & Dickson, F. W. 1975 *Earth planet. Sci. Lett.* **25**, 385–397.

Bischoff, J. L. & Sayles, F. L. 1972 *J. Sedim. Petrol.* **42**, 711–724.

Bjornsson, S., Arnorsson, S. & Tomasson, J. 1972 *Bull. Am. Ass. Petrol. Geol.* **56**, 2380–2391.

Bonatti, E. 1975 *A. Rev. Earth planet. Sci.* **3**, 401–431.

Bonatti, E., Honnorez-Guerstein, M. B., Honnorez, J. & Stern, C. 1976 *Earth planet. Sci. Lett.* **32**, 1–10.

Bonatti, E. & Joensuu, O. 1966 *Science, N.Y.* **154**, 643–645.

Boström, K. 1967 In *Researches in geochemistry* (ed. P. H. Abelson), pp. 421–452. New York: Wiley.

Boström, K. 1970 *Earth planet. Sci. Lett.* **9**, 348–354.

Boström, K., Farquharson, B. & Eyl, W. 1971 *Chem. Geol.* **10**, 189–203.

Boström, K. & Fisher, D. E. 1971 *Earth planet. Sci. Lett.* **11**, 95–98.

Boström, K. & Peterson, M. N. A. 1966 *Econ. Geol.* **61**, 1258–1265.

Boström, K. & Peterson, M. N. A. 1970 *Mar. Geol.* **7**, 427–447.

Boström, K., Peterson, M. N. A., Joensuu, O. & Fisher, D. E. 1969 *J. geophys. Res.* **74**, 3261–3270.

Boulad, A. P., Condomines, M., Bernat, M., Michard, G. & Allègre, C. 1975 *C.r. hebd. Séanc. Acad. Sci., Paris* D **280**, 2425–2428.

Bricker, O. 1965 *Am. Miner.* **50**, 1296–1354.

Burnett, W. C. & Morgenstein, M. 1976 *Earth planet. Sci. Lett.* **33**, 208–218.

Burnett, W. C. & Piper, D. Z. 1977 *Nature, Lond.* **265**, 596–600.

Burns, R. G. 1965 *Nature, Lond.* **205**, 999–1001.

Burns, R. G. 1976 *Geochim. cosmochim. Acta* **40**, 95–102.

Burns, R. G. & Brown, B. A. 1972 In *Ferromanganese deposits of the ocean floor* (ed. D. R. Horn), pp. 51–60. Washington, D.C.: National Science Foundation.

Burns, R. G. & Burns, V. M. 1975 *Nature, Lond.* **255**, 130–131.

Burns, R. G. & Burns, V. M. 1977 In *Marine manganese deposits* (ed. G. P. Glasby), pp. 185–248. Amsterdam: Elsevier.

Burns, R. G. & Fuerstenau, D. W. 1966 *Am. Miner.* **51**, 895–901.

Buser, W. 1959 In *International Oceanography Congress*, reprints, pp. 962–963. Amer. Assoc. Adv. Science.

Buser, W. & Graf, P. 1955 *Helv. chim. Acta* **38**, 830–834.

Buser, W., Graf, P. & Feitknecht, W. 1954 *Helv. chim. Acta* **37**, 2322–2333.

Buser, W. & Grütter, A. 1956 *Schweiz. miner. Petrogr. Mitt.* **36**, 49–62.

Byrne, R. H. & Kester, D. R. 1976 *Mar. Chem.* **4**, 255–274.

Calvert, S. E. & Price, N. B. 1970 *Contr. Miner. Petr.* **29**, 215–233.

Calvert, S. E. & Price, N. B. 1977a *Mar. Chem.* **5**, 43–74.

Calvert, S. E. & Price, N. B. 1977b In *Marine manganese deposits* (ed. G. P. Glasby), pp. 45–86. Amsterdam: Elsevier.

Calvert, S. E., Price, N. B., Heath, G. R. & Moore, T. C. 1978 *J. mar. Res.* (In the press.)

Cheney, E. S. & Vredenburgh, L. D. 1968 *J. Sedim. Petrol.* **38**, 1363–1365.

Chukhrov, F. V., Zvyagin, B. B., Gorshkov, A. I., Yermilova, L. P. & Balashova, V. V. 1973 *Int. Geol. Rev.* **16**, 1131–1143.

Coey, J. M. D. & Readman, P. W. 1973 *Earth planet. Sci. Lett.* **21**, 45–51.

Cook, H. E. 1971 *Geol. Soc. Am. Abstr.* **3**, 530–531.

Corliss, J. B. 1971 *J. geophys. Res.* **76**, 8128–8138.

Crerar, D. A. & Barnes, H. L. 1974 *Geochim. cosmochim. Acta* **38**, 279–300.

Cronan, D. S. 1967 Ph.D. thesis, University of London, 342 pages.

Cronan, D. S. 1972 *Can. J. Earth Sci.* **9**, 319–323.

Cronan, D. S. 1973 In *Initial reports of the Deep Sea Drilling Project* (eds T. H. van Andel *et al.*), vol. 16, pp. 601–604. Washington, D.C.: U.S. Government Printing Office.

Cronan, D. S. 1974 In *The sea* (ed. E. D. Goldberg), vol. 5, pp. 491–525. New York: Wiley.

Cronan, D. S. 1975 *J. Geophys. Res.* **80**, 3831–3837.

Cronan, D. S. 1976a In *Chemical oceanography*, 2nd edn (eds J. P. Riley and R. Chester), vol. 5, pp. 217–263. London: Academic Press.

Cronan, D. S. 1976b *Bull. geol. Soc. Am.* **87**, 928–934.

Cronan, D. S., van Andel, T. H., Heath, G. R., Dinkelman, M. G., Rodolfo, K. S., Yeats, R. S., Bennett, R. H., Bukry, D., Charleston, S. & Kaneps, A. 1972 *Science, N.Y.* **175**, 61–63.

Cronan, D. S. & Garrett, D. E. 1973 *Nature, phys. Sci.* **242**, 88–89.

Cronan, D. S. & Tooms, J. S. 1967 *Deep Sea Res.* **14**, 117–119.

Cronan, D. S. & Tooms, J. S. 1968 *Deep Sea Res.* **15**, 215–223.

Cronan, D. S. & Tooms, J. S. 1969 *Deep Sea Res.* **16**, 335–359.

Dasch, E. J. 1969 *Geochim. cosmochim. Acta* **33**, 1521–1522.

Dasch, E. J., Dymond, J. R. & Heath, G. R. 1971 *Earth planet. Sci. Lett.* **13**, 175–180.

Degens, E. T. & Ross, D. A. 1969 *Hot brines and recent heavy metal deposits in the Red sea.* New York: Springer-Verlag, 600 pages.

Detrick, R. S., Williams, D. L., Mudie, J. D. & Sclater, J. G. 1974 *Geophys. J. R. astr. Soc.* **38**, 627–637.

Drever, J. I. 1971 In *Initial Reports of the Deep Sea Drilling Project* (eds E. L. Winterer *et al.*), vol. 7, part 2, pp. 965–975. Washington, D.C.: U.S. Government Printing Office.

Dunham, A. C. & Glasby, G. P. 1974 *N.Z. J. Geol. Geophys.* **17**, 929–953.

Dymond, J. R., Corliss, J. B., Heath, G. R., Field, C. W., Dasch, E. J. & Veeh, H. H. 1973 *Bull. geol. Soc. Am.* **84**, 3355–3372.

Dymond, J. & Veeh, H. H. 1975 *Earth planet. Sci. Lett.* **28**, 13–22.

Ehrlich, H. L. 1964 In *Principles and applications in aquatic microbiology* (eds H. Heukelekian & N. C. Dondero) pp. 43–60. New York: Wiley.

Ehrlich, H. L. 1966 *Devs ind. Microbiol.* **7**, 279–286.

Eklund, W. A. 1974 M.S. thesis, Oregon State University, 77 pages.

Elderfield, H. 1976 *Mar. Chem.* **4**, 103–132.

Elderfield, H., Gunnlaugsson, E., Wakefield, S. J. & Williams, P. T. 1977 *Miner. Mag.* **41**, 217–226.

Ellis, A. J. 1970 In *Proc. 9th Commonwealth Mining and Metallurgical Congress*, vol. 2, pp. 211–240. London: Inst. Mining and Metallurgy.

Ellis, A. J. 1973 In *Proc. Symp. Hydrogeochem. and Biogeochem.* (ed. E. Ingerson), pp. 1–26. Washington, D.C.: The Clarke Co.

El-Sayed, S. Z. 1970 In *Scientific exploration of the South pacific* (ed. W. S. Wooster), pp. 194–210. Washington, D.C.: Natn. Acad. Sci.

El Wakeel, S. K. & Riley, J. P. 1961 *Geochim. cosmochim. Acta* **25**, 110–146.

Fisher, D. E. & Boström, K. 1969 *Nature, Lond.* **224**, 64–66.

Frey, F. A. & Haskin, L. A. 1968 *J. geophys. Res.* **69**, 775–785.

Friedrich, G., Rosner, B. & Demirsoy, S. 1969 *Miner. Deposita* **4**, 298–307.

Froelich, P. N., Bender, M. L. & Heath, G. R. 1977 *Earth planet. Sci. Lett.* **34**, 351–359.

Giovanoli, R. & Bürki, P. 1975 *Chimia* **29**, 266–269.

Giovanoli, R., Burki, P., Gluffredi, M. & Stumm, W. 1975 *Chimia* **29**, 517–520.

Glasby, G. P. 1971 *N.Z. J. Sci.* **15**, 232–239.

Glasby, G. P. 1972*a* *Mar. Geol.* **13**, 57–72.

Glasby, G. P. 1972*b* *Nature phys. Sci.* **237**, 85–86.

Glasby, G. P. 1972*c* *N.Z. J. Sci.* **15**, 232–239.

Glasby, G. P. 1976 *N.Z. J. Geol. Geophys.* **19**, 707–736.

Glasby, G. P. (ed.) 1977 *Marine manganese deposits.* (523 pages.) Amsterdam: Elsevier.

Glasby, G. P., Bäcker, H. & Meylan, M. A. 1975 *Erzmetall.* **28**, 340–342.

Glasby, G. P., Tooms, J. S. & Howarth, R. J. 1974 *N.Z. J. Sci.* **17**, 387–407.

Glemser, O., Gattow, G. & Meisiek, H. 1961 *Z. anorg. allg. Chem.* **309**, 1–19.

Goldberg, E. D. 1954 *J. Geol.* **62**, 249–265.

Goldberg, E. D. 1961 In *Oceanography* (ed. M. Sears), pp. 583–597. (Am. Assoc. Adv. Sci. Publ. 67.)

Goldberg, E. D. & Arrhenius, G. 1958 *Geochim. cosmochim. acta* **13**, 153–212.

Goldberg, E. D., Koide, M., Schmitt, R. & Smith, R. 1963 *J. geophys. Res.* **68**, 4209–4217.

Goodell, H. G., Meylan, M. A. & Grant, B. 1971 *Antarctic Research Series* **15**, 27–92.

Greenslate, J. 1974*a* *Nature, Lond.* **249**, 181–183.

Greenslate, J. 1974*b* *Science, N.Y.* **186**, 529–531.

Greenslate, J. L., Frazer, J. Z. & Arrhenius, G. 1973 In *Papers on the origin and distribution of manganese nodules in the Pacific and prospects for exploration* (ed. M. Morgenstein), pp. 45–69. Honolulu, Hawaii: Valdivia Manganese Exploration Group, University of Hawaii and National Science Foundation.

Grütter, A. & Buser, W. 1957 *Chimia* **11**, 132–133.

Hajash, A. 1975 *Contr. Miner. Petr.* **53**, 205–226.

Hart, S. R. 1971 *Phil. Trans. R. Soc. Lond.* A **268**, 573–587.

Hart, R. A. 1973 *Can. J. Earth Sci.* **10**, 799–815.

Heath, G. R. & Dymond, J. 1977 *Bull. geol. Soc. Am.* **88**, 723–733.

Hem, J. D. 1977 *Geochim. cosmochim. Acta* **41**, 527–539.

Heye, D. 1969 *Earth planet. Sci. Lett.* **6**, 112–116.

Heye, D. 1975 *Geol. Jb.* E **5**, 3–122.

Heye, D. & Beiersdorf, H. 1973 *Z. Geophys.* **39**, 703–726.

Heye, D. & Marchig, V. 1977 *Mar. Geol.* **23**, M19–M25.

Hildreth, R. A. & Henderson, W. T. 1971 *Geochim. cosmochim. Acta* **35**, 235–238.

Høgdahl, O. T., Melsom, S. & Bowen, V. T. 1968 *Adv. Chem. Ser.* **73**, 308–325.

Horn, D. R., Delach, M. N. & Horn, B. M. 1973 *Metal content of ferromanganese deposits of the oceans*. I.D.O.E. Technical Report No. 3. National Science Foundation, Washington, D.C. 51 pages (unpublished).

Horn, M. K. & Adams, J. A. S. 1966 *Geochim. cosmochim. Acta* **30**, 279–297.

Horowitz, A. 1970 *Mar. Geol.* **9**, 241–259.

Jenkyns, H. C. & Hardy, R. G. 1976 In *Initial reports of the Deep-Sea Drilling Project* (eds S. O. Schlanger *et al.*), vol. 33, pp. 833–836. Washington: U.S. Government Printing Office.

Joides 1967 *Bull. Am. Ass. Petrol. Geol.* **51**, 1787–1802.

Jones, L. H. P. & Milne, A. A. 1956 *Miner. Mag.* **31**, 283–288.

Kazachevskii, I. V., Cherdyntsev, V. V., Kua'Mina, E. A., Sulerzhitskii, L. D., Mochalova, V. F. & Kyuregyan, T. N. 1964 *Geokhimiya*, no. 11, 1116–1121.

Klinkhammer, G., Weiss, R. F. & Bender, M. 1977 *Nature, Lond.* **269**, 319–320.

Koblentz-Mishke, O. J., Volkovinsky, V. V. & Kabanova, J. G. 1970 In *Scientific exploration of the South Pacific* (ed. W. S. Wooster), pp. 183–193. Washington, D.C.: National Academy of Sciences.

Kraemer, T. & Schornick, J. C. 1974 *Chem. Geol.* **13**, 187–196.

Krishnaswami, S. (1976) *Geochim. cosmochim. Acta* **40**, 425–434.

Krishnaswami, S. & Lal, D. 1972 In *Nobel Symposium 20: the changing chemistry of the oceans* (eds D. Dyrssen & D. Jagner), pp. 307–319. New York: Wiley.

Ku, T. L. & Broecker, W. S. 1969 *Deep Sea Res.* **16**, 625–637.

Ku, T. L., Broecker, W. S. & Opdyke, N. 1968 *Earth planet. Sci. Lett.* **4**, 1–16.

Kunzendorf, H. & Friedrich, G. H. W. 1976 *Geochim. cosmochim. Acta* **40**, 849–852.

Lalou, C. & Brichet, E. 1972 *C.r. hebd. Séanc. Acad. Sci., Paris* D **275**, 815–818.

Lalou, C., Brichet, E. & L. E. Gressus, C. 1973*a* *Annls. Inst. oceanogr., Paris* **49**, 5–17.

Lalou, C., Brichet, E & Ranque, D. 1973*b* *C.r. hebd. Séanc. Acad. Sci., Paris* D **276**, 1661–1664.

Lalou, C., Delibrias, G., Brichet, E. & Labeyrie, J. 1973*c* *C.r. hebd. Séanc. Acad. Sci, Paris* D **276**, 3013–3015.

Lister, C. R. B. 1972 *Geophys. J. R. astr. Soc.* **26**, 515–535.

Lupton, J. E., Weiss, R. F. & Craig, H. 1977 *Nature, Lond.* **267**, 603–604.

Lyle, M. 1976 *Geology* **4**, 733–736.

Lyle, M., Dymond, J. & Heath, G. R. 1977 *Earth planet. Sci. Lett.* **35**, 55–64.

Manheim, F. T. 1965 *Manganese–iron accumulations in the shallow-water environment*. Narragansett Marine Laboratory University of Rhode Island, Occas. Publ. No. 3, pp. 217–276.

Manheim, F. T. 1974 In *Initial reports of the Deep Sea Drilling Project* (ed. R. B. Whitmarsh), vol. 23, pp. 975–998. Washington, D.C.: U.S. Government Printing Office.

Margolis, S. V. & Burns, R. G. 1976 *A. Rev. Earth planet. Sci.* **4**, 229–263.

Margolis, S. V. & Glasby, G. P. 1973 *Bull. geol. Soc. Am.* **84**, 3601–3610.

McMurtry, G. M. 1975 Hawaii Inst. Geophys. Publ. HIG-75-14, pp. 1–40.

Mero, J. L. 1962 *Econ. Geol.* **57**, 747–767.

Mero, J. L. 1965 *The mineral resources of the sea*. Amsterdam: Elsevier, 312 pages.

Meylan, M. A. & Goodell, H. G. 1976 In *United Nations Economic and Social Commission for Asia and the Pacific. Technical Bulletin No. 2* (eds G. P. Glasby & H. R. Katz), pp. 99–117.

Monty, C. 1973 *C.r. hebd. Séanc. Acad. Sci., Paris* D **276**, 3285–3288.

Moore, W. S. & Vogt, P. R. 1976 *Earth planet. Sci. Lett.* **29**, 349–356.

Mottl, M. J., Corr, R. F. & Holland, H. D. 1974 *Geol. Soc. Am. Abstr.* **6**, 879–880.

Murray, J. & Renard, A. F. 1891 *Deep sea deposits. Repts. Scientific Results Explor. Voyage H.M.S. Challenger, 1873–1876*. H.M.S.O. 525 pages.

Nikolayev, D. S. & Yefimova, E. I. 1963 *Geokhimiya*, no. 7, pp. 703–714.

Otswald, J. & Fraser, F. W. 1973 *Miner. Deposita* **8**, 303–311.

Perfil'ev, B. V., Gabe, D. R., Gal'Pernia, A. M., Rabinovich, V. A., Sapotnitskii, A. A., Sherman, E. E. & Troshanov, E. P. 1965 *Applied capillary microscopy*. New York: Consultants Bureau, 122 pages.

Peterson, M. N. A. 1966 *Science, N.Y.* **154**, 1542–1544.

Piper, D. Z. 1973 *Earth Planet. Sci. Lett.* **19**, 75–82.

Piper, D. Z. 1974*a* *Geochim. cosmochim. Acta* **38**, 1007–1022.

Piper, D. Z. 1974*b* *Chem. Geol.* **14**, 285–304.

Piper, D. Z. & Graef, P. A. 1974 *Mar. Geol.* **17**, 287–297.

Piper, D. Z., Veeh, H. H., Bertrand, W. G. & Chase, R. L. 1975 *Earth planet. Sci. Lett.* **26**, 114–120.

Piper, D. Z. & Williamson, M. E. 1977 *Mar. Geol.* **23**, 285–303.

Price, N. B. & Calvert, S. E. 1970 *Mar. Geol.* **9**, 145–171.

Raab, W. 1972 In *Ferromanganese deposits on the ocean floor* (ed. D. R. Horn), pp. 31–50. Washington, D.C.: National Science Foundation.

Reid, J. L. 1962 *Limnol. Oceanog.* **7**, 287–306.

Revelle, R. R. 1944 Marine bottom samples collected in the Pacific Ocean by the Carnegie on its seventh cruise. *Publ. Carnegie Instn. Wash.* **556**, pp. 1–180.

Reynolds, P. H. & Dasch, E. J. 1971 *J. geophys. Res.* **76**, 5124–5129.

Riley, J. P. & Sinhaseni, P. 1958 *J. Mar. Res.* **17**, 466–482.

Rona, P. A., McGregor, B. A., Betzer, P. R., Bolger, G. W. & Krause, D. C. 1975 *Deep Sea Res.* **22**, 611–618.

Rona, P. D., Harbison, R. N., Bassinger, B. G., Scott, R. B. & Nalwalk, A. J. 1976 *Bull. geol. Soc. Am.* **87**, 661–674.

Rydell, S. & Bonatti, E. 1973 *Geochim. cosmochim. Acta* **37**, 2557–2565.

Sayles, F. L. & Bischoff, J. L. 1973 *Earth planet. Sci. Lett.* **19**, 330–336.

Sayles, F. L., Ku, T.-L. & Bowker, P. C. 1975 *Bull. geol. Soc. Am.* **86**, 1423–1431.

Scott, R. B., Malpas, J., Rona, P. A. & Udintsev, G. 1976 *Geology* 4, 233–236.

Scott, R. B., Rona, P. A., Butler, L. W., Nalwalk, A. J. & Scott, M. R. 1972 *Nature, phys. Sci.* **239**, 77–79.

Scott, R. B., Rona, P. A., McGregor, B. A. & Scott, M. R. 1974*b Nature, Lond.* **251**, 301–302.

Scott, M. R., Scott, R. B., Rona, P. A., Butler, L. W. & Nalwalk, A. J. 1974 *Geophys. Res. Lett.* **1**, 355–358.

Scott, M. R., Scott, R. B., Nalwalk, A. J., Rona, P. A. & Butler, L. W. 1973 *Trans. Am. geophys. Un.* **54**, 244.

Scott, R. B., Scott, M. R., Swanson, S. B., Rona, P. A. & McGregor, B. A. 1974*a Trans. Am. geophys. Un.* **55**, 293.

Sillén, L. G. 1961 In *Oceanography* (ed. M. Sears), pp. 549–581 (Am. Assoc. Adv. Sci. Publ. 67).

Skornyakova, N. S. 1964 *Litol. i Polezn. Iskop.* no. 5, 3–20.

Skornyakova, N. S. 1976 *Trans. P.P. Shirshov Inst. Oceanol.* **109**, 190–240.

Skornyakova, N. S. & Andrushcheuko, P. F. 1972 *Int. geol. Rev.* **16**, 863–919.

Skornyakova, N. S., Bazilevskaya, E. S. & Gordevev, V. V. 1975 *Geokhimiya* No. 7, 1064–1076.

Somayajulu, B. L. K. 1967 *Science, N.Y.* **156**, 1219–1220.

Somayajulu, B. L. K., Heath, G. R., Moore, T. C., Jr & Cronan, D. S. 1971 *Geochim. cosmochim. Acta* **35**, 621–624.

Sorem, R. K. 1967 *Econ. Geol.* **62**, 141–147.

Sorem, R. K. & Foster, A. R. 1972 In *Ferromanganese deposits on the ocean floor* (ed. D. R. Horn), pp. 167–179. Washington, D.C.: National Science Foundation.

Sorokin, Y. I. 1972 *Oceanology* **12**, 1–11.

Spooner, E. T. C. & Fyfe, W. S. 1973 *Contr. Miner. Petr.* **42**, 287–304.

Strakhov, N. M. 1974*a Litol. i Polezn. Iskop.* no. 5, 3–17.

Strakhov, N. M. 1974*b Litol. i Polezn. Iskop.* no. 3, 20–37.

Stumm, W. & Morgan, J. J. 1970 *Aquatic chemistry.* New York: Wiley, 583 pages.

Sugimura, Y., Miyake, Y. & Yanagawa, H. 1975 *Pap. Met. Geophys., Tokyo* **26**, 47–54.

Tatsumoto, M. 1966 *Science, N.Y.* **153**, 1094–1101.

Thurber, D. L. 1962 *J. geophys. Res.* **67**, 4518–4520.

Towe, K. M. & Bradley, W. F. 1967 *J. Colloid Interface Sci.* **24**, 384–392.

van der Weijden, C. H. 1976 *Chem. Geol.* **18**, 65–80.

Varentsov, I. M. 1971 *Soc. Mining Geol. Japan, Spec. Issue* **3**, 466–473.

Veeh, H. H. & Bostrom, K. 1971 *Earth planet. Sci. Lett.* **10**, 372–374.

von der Borch, C. C. & Rex, R. W. 1970 In *Initial reports of the Deep Sea Drilling Project* (ed. D. A. McManus *et al.*), vol. 5, pp. 541–544. Washington, D.C.: U.S. Government Printing Office.

von der Borch, C. C., Nesteroff, W. D. & Galehouse, J. S. 1971 In *Initial reports of the Deep Sea Drilling Project* (eds J. I. Tracey, Jr *et al.*), vol. 8, pp. 829–836. Washington, D.C.: U.S. Government Printing Office.

Wedepohl, K. 1960 *Geochim. cosmochim. Acta* **18**, 200–231.

Weiss, R. F., Lonsdale, P. F., Lupton, J. E., Bainbridge, A. E. & Craig, H. 1977 *Nature, Lond.* **267**, 600–603.

Wendt, J. 1974 *Spec. Publ. int. Ass. Sedimentol.* **1**, 437–447.

Williams, D. L. & von Herzen, R. P. 1974 *Geology* **2**, 327.

Willis, J. P. 1970 M.Sc. Thesis, University of Cape Town, 110 pages.

Willis, J. P. & Ahrens, L. H. 1962 *Geochim. cosmochim. Acta* **26**, 751–764.

Wolery, T. J. & Sleep, N. H. 1976 *J. Geol.* **84**, 249–275.

Yoshimura, T. 1934 *J. Fac. Sci. Hokkaido Univ.*, series 4, **2**, 289–297.

Zelenov, K. K. 1964 *Dokl. Akad. Nauk. S.S.S.R.* **155**, 1317–1320.

Discussion

D. S. CRONAN (*Geology Department, Imperial College, London S.W.7, U.K.*). I should like to amplify the reference Dr Calvert made to current work of our group at Imperial College on manganese nodules from the Indian Ocean. We have analysed samples of nodules and encrustations from more than sixty Indian Ocean sites. The results of these analyses coupled with previously published data have shown the existence of an area in the central south equatorial part of the ocean where nodules are very similar in composition to those in the northeastern equatorial Pacific 'ore zone' described by Dr Calvert in that they are rich in Mn, Ni, Cu and Zn (Cronan & Moorby 1976). The sparse data that we have been able to obtain so far on the conditions under which these nodules are forming indicate that they may be similar to those under which the high-grade nodules in the northeastern equatorial Pacific are deposited. The nodules are situated at the margin of the equatorial zone; the few available data on which suggest it to be an area of high biological productivity; their environment of deposition leads to todorokite being their principal mineral phase; and they are often found in association with siliceous sediments. The occurrence of these deposits under conditions seemingly similar to those under which high grade nodules occur in the Pacific should allow us to test the hypotheses which have been proposed to explain the origin of the latter. Furthermore, ongoing work on these deposits will not only help to outline an area of potential mining interest in the Indian Ocean, but should also help in the development of a predictive model which could be used to outline other possible areas of high-grade nodules.

Reference

Cronan, D. S. & Moorby, S. A. 1976 In *Marine geological investigations in the southwest Pacific and adjacent areas* (eds G. P. Glasby & H. R. Katz), U.N. Economic and Social Commission for Asia and the Pacific, Tech. Bull. no. 2, pp. 118–123.

Phil. Trans. R. Soc. Lond. A. **290**, 75–85 (1978)   [75]

Printed in Great Britain

Morphology of the continental margin

By A. S. Laughton and D. G. Roberts

Institute of Oceanographic Sciences, Wormley, Godalming, Surrey GU8 5UB, U.K.

[Plate 1; pullouts 1 and 2]

The continental margin is the surface morphological expression of the deeper fundamental transition between the thick low density continental igneous crust and the thin high density and chemically different oceanic igneous crust. Covering the transition are thick sediment accumulations comprising over half the total sediments of the ocean, so that the precise morphological boundaries often differ in position from those of the deeper geology.

Continental margins are classified as active or passive depending on the level of seismicity. Active continental margins are divided into two categories, based on the depth distribution of earthquakes and the tectonic régime. Active transform margins, characterized by shear and shallow focus earthquakes, result from horizontal shear motion between plates. Active compressional margins are characterized by shallow, intermediate and deep earthquakes along a dipping zone, by oceanic trenches and by volcanic island arcs or mountain ranges depending on whether the margin is ocean–ocean or ocean–continent.

Passive margins, found in the Atlantic and Indian Oceans, are formed initially by the rifting of continental crust and mark the ocean–continent boundary within the spreading plate. They are characterized by continental shelf, slope and rise physiographic provinces. Once clear of the rifting axis, they cool and subside. Sedimentation can prograde the shelf and load the edge leading to further downwarping; changes of sea level lead to erosion by wave action and by ice; ocean currents and turbidity currents redistribute sediments; slumps occur in unstable areas.

The passive and sediment-starved margin west of Europe is described where the following factors have been significant: (*a*) faulting related to initial rifting; (*b*) infilling and progradation by sediments; (*c*) slumping; (*d*) contour current erosion and deposition; (*e*) canyon erosion.

Introduction

The continental margins of the world, comprising the continental shelves, slopes and rises, cover 74×10^6 km², or a total of 15 % of the Earth's surface (Menard & Smith 1966). In the context of this discussion meeting it is neither possible nor desirable to review comprehensively all the complexities of margin morphology. It is more appropriate to consider those aspects of morphology that are likely to be relevant to the needs of ocean engineers as the exploitation of natural resources moves progressively from the shelf into deeper water.

Rather than limit the discussion to a straight description of the shape, features and texture of the margin, we review the underlying geological processes which have moulded the margins, with especial emphasis on those of the northeast Atlantic, because of the national interest. It is convenient to consider first the initial evolution of the margins and the development of the large-scale morphology, and then to examine the secondary processes that have sculpted the intermediate and small-scale features.

EVOLUTION OF THE LARGE SCALE MORPHOLOGY OF MARGINS

Plate tectonics (Le Pichon, Francheteau & Bonnin 1973) can account very satisfactorily for the splitting and separation of continents and the evolution and destruction of oceanic crust† during the past 200 Ma. The continental (or oceanic) margins reflect the transition between the thin and dense igneous crust and the thicker, less dense and chemically different continental or intermediate crust. Isostatic balance between the two crusts, modified by subsequent sedimentation or erosion, creates a major step in the seabed at the boundary, of which the continental slope is the prime expression.

Continental margins can be divided into passive and active types based on their seismicity. Passive margins lack earthquakes and widespread volcanism, unless very young, and are typically comprised of a continental shelf, slope and rise. In contrast, active margins like those that border the Pacific Ocean are associated with a trench, volcanism, mountain building and earthquakes that extend down to a depth of as much as 700 km along a dipping zone. The active margins are often associated with island arcs, marginal seas and interarc basins.

The division into active and passive types is a fundamental one that can best be understood in terms of the plate tectonics hypothesis in which the tectonic and seismic activity of the upper layer of the Earth is related to the interaction of a number of large rigid plates whose boundaries are the seismic belts of the world. These plates are diverging, converging or shearing past each other.

Active continental margins mark the boundaries between two plates which are either converging or sliding past each other with deformation or destruction of crust, often by subduction back into the interior of the Earth. By contrast, the passive margins are the ancient scars of the rifting and subsequent separation of the halves of a continental plate and which have migrated away from the active zone as new oceanic crust fills the gap.

This paper is concerned with reviewing the evolution of passive margins which are found throughout nearly all the Atlantic, and discussing the geological processes that have shaped their present morphology.

Passive continental margins

Although many passive continental margins appear to have formed by rifting, their present structure, stratigraphy and morphology result from a number of geological processes that are a function of climate and ocean circulation as well as being variable in time and space. A better understanding of the relationship of these factors within a kinematic framework of continental margin evolution is now developing as the first results in 1976 and 1977 of the I.P.O.D.–D.S.D.P.‡ passive margin drilling programme, and the accompanying geophysical surveys, become available. From these new data, and by comparison with present-day analogue margins, we can gain a perspective within which to review passive margin evolution (Falvey 1972; Roberts & Caston 1975). The highly schematic evolutionary sequence shown in figure 1 forms the basis of the following discussion.

The evolution of a passive margin is considered to include the following phases:
(1) rifting of the continental crust;

---

† The term 'crust' is sometimes used to include all rocks above the Mohorovičic discontinuity and is sometimes limited, as in this paper, to the igneous or crystalline rocks only.

‡ International Phase of Ocean Drilling of the Deep-Sea Drilling Project.

(2) onset of spreading (i.e. actual separation of the continental crust and the accretion of oceanic crust in the gap between the continental blocks);

(3) post-rift evolution (subsidence of the rifted margins and shaping of the present morphology by tectonic and sedimentary processes).

Evidence for the nature of the rifting process that constitutes the initial stage of margin development is based on the analogue provided by the East African rift system. A period of regional uplift may precede or be contemporaneous with rifting and accompanied by widespread volcanism. The rifting may follow subparallel or trilete patterns exemplified by the Viking Graben of the North Sea. During this stage, the basic structural framework of the

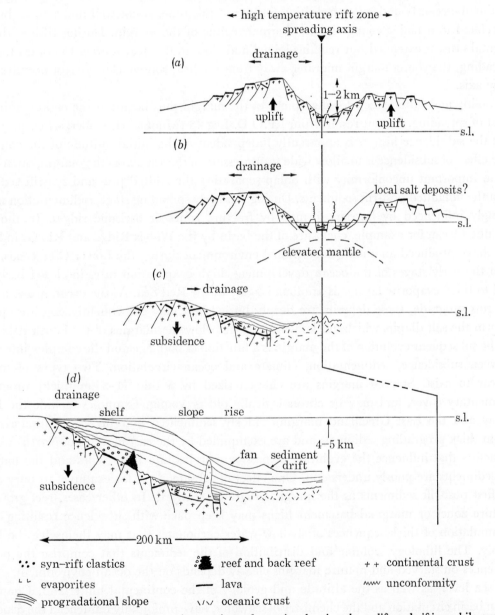

FIGURE 1. Stages in the evolution of a passive continental margin, showing the uplift and rifting while over the high temperature zone, and subsidence and sedimentation when it has moved away. (a) Rifting; (b) crustal attenuation; (c) initial spreading; (d) old margin. The line marked s.l. represents sea level.

embryo margin is shaped by the rifting and fracture zone development, which is influenced by the pre-existing structural fabric of the continent. Within the rift, basic and alkaline extrusives may be intercalated with, and thickly covered by, contemporaneous coarse clastic continental sediments. An effect of uplift is to direct the drainage away from the rift valley restricting the supply of clastic sediment. In other cases, the rifting may take place in an epicontinental sea, creating a deep (*ca.* 2 km) rift basin.

As the rift widens, the continental crust may stretch or fracture, and thus become thinner, allowing mantle rocks to rise. Eventually partial melts from the mantle penetrate the continental crust and oceanic crust begins to accrete at the edges of the separating blocks. The nature of the change from rifting to spreading and accretion, and thus the geology of the continent–ocean boundary or transition formed at this time, is not well understood, but may be related to a major change in the thermal régime of the margin. During rifting, the continental crust is extended but remains joined and close to the heat source. In contrast, during spreading, the young margin migrates away from the heat source as new crust accretes at the ridge axis.

Cooling and subsidence (downwarping) of the margin thus begins at the end of rifting and onset of spreading. Recent results from I.P.O.D. Leg 48 (Montadert, Roberts *et al.* 1977) show that the subsidence history is apparently independent of the initial altitude of the continent. One effect of subsidence is to allow wide transgression of the sea across the margin, often shown by an important unconformity with onlap separating the faulted pre- and syn-rift sediments from the unfaulted post-rift sediments. During this early spreading stage, sedimentation may be strongly influenced by barriers formed by fracture zones or aseismic ridges. In the south Atlantic Ocean for example, enclosure of the basin by the Walvis Ridge and Rio Grande Rise may have produced an anoxic depositional environment during the Lower Cretaceous.

In the early stages of the ocean development, high evaporation rates in closed basins can lead to thick evaporite layers, as are found today in the Red Sea. As the ocean grows and the margins prograde, these deposits are deeply buried and may subsequently migrate upwards to form the salt diapirs which are known to occur below the margins of the South Atlantic.

The subsequent evolution of the margin is a function of both age and the complex interaction between subsidence, sedimentation, climate and ocean circulation. Two types of margins appear to exist. Starved margins are characterized by a thin (1–2 km thick) prograding sedimentary cover, and may be chronologically old or young. Examples include the Bay of Biscay and the East Greenland margin. Thickly sedimented margins are characterized by 10 km thick prograding sediments and are exemplified by the east margins of North America.

Factors that influence the evolution of passive margins into these types and the nature of the sediments are poorly understood. One effect of subsidence is, for example, to bury deeply the first post-rift sediments as the margin progrades seaward. In other cases, reef growth on fracture zones or marginal basement highs may keep pace with subsidence resulting in the accumulation of thick sequences of shallow-water carbonates that may themselves be buried deeply. The lithology, volume and distribution of the sediments that comprise the post-rift sequences on starved and mature margins clearly depends on the ocean environment, climate and sea level, as well as the altitude and geology of the continental hinterland. Changes in ocean basin circulation and the chemistry of seawater, have profoundly influenced the deposition of sediments along continental margins producing large hiatuses in the geological record.

Although the integrated effects of these factors are extremely complex, there may be a few simple controls. For example, the lithology and volume of the sediments deposited on the margin may be strongly influenced by global changes in sea level. Such changes may reflect variations in spreading rate (Flemming & Roberts 1973) which, by altering the cross-sectional area of mid-ocean ridges and hence the ocean basin volume, may cause changes in ocean circulation and climate.

In summary, the gross morphology of the passive margins arises from (a) the initial rift and associated faulting and local sedimentation while the margin lay over the thermal zone and, (b) subsequent progradation, reefal growth, erosion, diapirism and subsidence as the margin cools and shrinks.

INTERMEDIATE AND SMALL-SCALE MORPHOLOGY

In discussing the intermediate and smaller-scale morphology it is convenient to distinguish between the classic subdivision of the margin (Heezen, Tharp & Ewing 1959) into shelf, slope and rise since the environments and processes operating in these regions differ significantly. It is important to remember, however, that on many margins these classifications cannot be applied too rigorously as there are sometimes deeply submerged and isolated continental blocks near the margin, multiple shelf–slope sequences, enclosed basins on the slope, and very often the continental rise is either absent as a morphological feature, or can only be delineated approximately (Emery 1971).

Continental shelf

The continental shelf has been extensively discussed in the literature and in many areas has been well surveyed using echo-sounding and side-scan sonar techniques. Its relatively flat surface arises from wave erosion during low sea-level stances during the last glacial period. The maximum lowering of between 70 and 140 m occurred about 15000 years ago (Emery, Niino & Sullivan 1971).

During glaciation, the land and shelf ice-sheets gouged great troughs to many hundreds of metres depth across the shelf. Although some of these troughs have been partly closed by moraines and infilled by later sediments, good examples are found along the Greenland, Labrador, Alaskan and Norwegian margins and in the North Sea as the Norwegian Trench (Holtedahl & Sellevoll 1971, 1972). Icebergs detached from the land or shelf ice have also been carried southward as far as 32° N by surface currents often to ground on the edge of the shelf and the upper slope ploughing furrows 10 m deep. Highly irregular criss-crossing patterns of iceberg plough-marks have been mapped by side-scan sonar along the shelf edge north and west of the U.K. and around the Rockall Bank further to the west (Belderson, Kenyon & Wilson 1973).

The present distribution and depositional régime of the shelf seas of NW Europe reflects the reworking of relict glacial sediments and the continued input of fresh sediments from the land areas. The bed load transport of these sediments has been related to the pattern of tidal currents by analysis of side-scan sonar records, but it is clear that wave action may also suspend the coarse fraction of the sediments (Kenyon & Stride 1970).

The continental slope and rise

Four major processes are responsible for fashioning the intermediate and small-scale morphology of the continental slope and rise: (a) pelagic sedimentation, (b) mass downslope transport of sediments by slumping, gravity slide or creep, (c) erosion of submarine canyons, and (d) the distribution or redistribution of sediments by deep contour-following currents.

(a) Pelagic sedimentation

Clay-sized sediments transported in suspension from the higher energy environment of the shelf to the quieter depths beyond are slowly deposited to mantle the slope and rise. These terrigeneous sediments are mixed with biogenic sediments largely composed of the calcareous and siliceous tests of the plankton populating the oceanic surface waters. Upwelling of nutrient-rich water over the continental slope may increase the plankton population and hence the proportion of biogenous sediments. Pelagic sediments cover by far the greater part of the continental slope and rise province which is characterized on the scale seen in seabed photographs by featureless unconsolidated sediments disturbed only by the burrows and tracks of benthic organisms. The cumulative effect of pelagic sedimentation is to subdue the relief of much of the slope and to contribute to the oceanward progradation of the shelf.

(b) Slumps, slides or creeps

Massive slumping and gravity sliding is now widely recognized as an important influence on the morphology and deposition of sediments in a variety of continental margin environments. The general condition for slumping of sediments deposited on a slope is that the shear strength of sediments along a potential glide plane must be exceeded by the shear stresses acting downslope. The shear strength is governed by sedimentation rate, grain size, lithology, age, degree of consolidation and the pore pressure conditions at the time of failure.

If the shear strength is exceeded gradually, the sediment may creep downslope with little visible disturbance of the surface except at the edges of the creeping mass. Slumps may occur on slopes oversteepened by deposition or may be caused by loading due, for example, to rapid deposition, previous slumping, or earthquake activity. In slumps due to collapse failure, the sediments are in a metastable condition caused primarily by loose packing of the sediments. Metastable accumulations of sediments often occur by continuous deposition at the heads of canyons or at the shelf break, though the metastability can be caused by wave action or by earthquakes.

A very large slump occurred in 1929 on the continental slope south of the Grand Banks off Newfoundland as a result of a nearby earthquake (Heezen & Drake 1964). The slump block, 400 m thick, 100 km wide and 100 km long, moved bodily downslope for about 100 m, breaking an array of trans-Atlantic submarine telegraph cables and generating turbidity currents that flowed many hundreds of kilometres into the ocean basins at speeds of up to 25 m s^{-1}, breaking further cables in succession (Heezen & Ewing 1952; Heezen, Ericson & Ewing 1954). Cable breaks in other areas have indicated similar recent slumps associated with either seismic activity (Heezen & Ewing 1955; Coulter & Migliaccio 1966), or with extreme river discharge (Heezen 1956). Less recent slumps have been identified from seismic profiles of the western side of Rockall Trough (Roberts 1972), and from near bottom observations of disturbed small-scale morphology in the adjacent trough (R. D. Flood, personal communications). Since this slump apparently originated on a slope as small as 2°, slumps may be a common occurrence on steeper slopes. The triggering mechanism for the Rockall slumps is not known but is thought to be associated with vigorous wave action during lowered Pleistocene sea level. If this is so, the slopes of Western Europe are likely to be more stable now than during the Pleistocene. Slump blocks may modify the morphology in such a way as to change significantly the course of adjacent canyons, or to create a barrier behind which sediments may be trapped.

Phil. Trans. R. Soc. Lond. A, volume 290

Laughton & Roberts, plate 1

FIGURE 2. Sonograph views of canyons near the top of the continental slope on the Armorican margin (Belderson & Kenyon 1976). The 13 km (*a*) and 7 km (*b*) views are obtained by insonification downslope with a narrow beam sonar. Highlights indicate strong echoes. (*a*) Slope at $45\frac{1}{2}°$ N, 4° W intensely eroded by canyons. Length of trace, 54 km. (*b*) Slope at $48\frac{1}{2}°$ N, $9\frac{1}{2}°$ W with few canyons. Length of trace, 32 km.

(c) Submarine canyon erosion

Submarine canyons are the most easily recognizable and most studied features of the continental slope (Shepard & Dill 1966; Shepard 1972). In places, canyon erosion dominates the morphology of the slope whereas in others, canyons are either isolated features or entirely absent. Many mechanisms have been proposed for their origin and maintenance, none of which apply universally. It is now generally agreed that submarine erosion by downslope sediment movement, either gradual in the form of bedcreep or violent as slumps and turbidity currents, is responsible for cutting and maintaining them, although the heads of some canyons which reach near the coast, such as Cap Breton Canyon in SE Biscay, may have been partly cut subaerially during lowered sea level. Some indeed, such as the Hudson Canyon east of the Hudson river, were once the outlet of present rivers.

Most canyons run normal to the continental slope, but many are deflected along structural boundaries such as the upper side of rotated slump blocks, faults (Boillot *et al.* 1974), or along the more easily eroded strata of the underlying rock. Accurate mapping of the course of canyons can reveal something of the underlying geological structure. Acoustic pictures of canyons north of Biscay (figure 2, plate 1) obtained with long range side-scan sonar (Belderson & Kenyon 1976), show several canyons with axes oblique to the regional slope which are possibly fault controlled, as well as features interpreted as slumps. These sonographs also revealed sets of secondary gullies on the sharp crested ridges between canyons often joining the canyon at angles near 90°. Sometimes these secondary gullies are left hanging above the axial gorge of the main canyon.

At the base of the continental slope, canyons debouch on to cones or fans where the rapid decrease in slope, and hence in speed of the transport mechanism, allows deposition of the sediments. On margins with many canyons, these cones coalesce into the apron of the continental rise. Submarine channels linked to the foot of canyons often are found crossing the rise and meandering across the relatively small slopes towards the ocean deeps. The canyons are usually a few tens of metres deep, a few hundred metres across and are sometimes leveed, indicating that they are caused by a sediment-laden density current mainly confined to the channel but occasionally overflowing its banks. The mechanism which allows these flows to continue for thousands of miles, as in the cases of the northwest Atlantic mid-ocean canyon (Heezen, Johnson & Hollister 1969) and the Bay of Bengal canyons (Curray & Moore 1971), is still not understood.

The small-scale morphology of the canyons is a direct result of the erosion mechanisms operating in them. Canyons which deeply incise the shelf edge can trap the coarser sediments being moved along the seabed under the action of currents. The resulting sand and gravel bodies in the canyons can creep or slump downslope along the canyon axis eroding, and even undercutting, the walls of the canyon, giving rise to steep-sided gorges (Shepard 1972), which have been explored in the top 100 m by divers and have been shown by deep submersible observations to be common also at great depths. Well sorted sands have been seen flowing as 'sand falls', akin to waterfalls, and must contribute to erosion and transport (Shepard & Dill 1966).

More violent, but spasmodic, events are turbidity currents which develop when a slump or other disturbance creates a body of dense sediment-laden water that flows downhill, gaining momentum, gaining more sediment by erosion and hence growing in density and in speed.

Speeds of many tens of metres per second have been observed for turbidity events when they reach the bottom of the slope and they can spread out hundreds of kilometres over the abyssal plains. The frequency of such events is difficult to estimate. However, the sedimentary record of the abyssal plains (e.g. D.S.D.P. Hole 118 in Laughton, Berggren *et al.* 1972) shows that they may occur at intervals between a few thousand and tens of thousands of years. Turbidity currents may be more common during lower sea level when the heads of the canyons were in a more energetic environment.

Clearly such violent processes deeply incise and erode canyons sweeping all loose sediments downslope. Underwater photographs have shown talus slopes of collapsed walls of canyons when they have been undercut by such a mechanism.

On a smaller scale, but perhaps as important in their cumulative effect, are the alternating up- and down-currents which have been observed in canyons down to depths of several thousand metres (Shepard, Marshall & McLoughlin 1974). At depths of the order of several hundred metres these have periodicities from 20 minutes to several hours and are believed to be related to internal waves (Shepard 1975), whereas in progressively deeper water the periodicity approaches that of the semidiurnal tide (Shepard 1976). Current velocities of up to 30 cm s^{-1} have been measured which are capable of transporting sediment, and the data show that there is a net down-canyon flow of water and of sediment. Underwater photographs show sand ripples indicative of sediment transport.

(d) Contour current sedimentation

Over the past decade, many studies have shown that deep ocean currents play an important rôle in fashioning, by erosion and deposition, a characteristic relief on the continental slope and rise (Hollister & Heezen 1972). In the northern hemisphere the rotation of the Earth deflects thermohaline ocean currents towards the right so that they tend to flow parallel and adjacent to contours of the major topographic features such as the continental slope. These currents are competent to transport fine sediment in suspension and sometimes to erode. These sediments are re-deposited as constructional ridges parallel to the current. Off the east coast of the U.S.A., for example, the Blake–Bahama Outer Ridge was formed by deposition from the Western Boundary Undercurrent (Heezen, Hollister & Ruddiman 1966). In the northeast Atlantic, comparable sediment ridges, some hundreds of kilometres in length and a few hundreds of metres in height, have been built up south of Greenland, east of the Reykjanes Ridge, and in the Rockall Trough (Johnson & Schneider 1969; Jones, Ewing, Ewing & Eittreim 1970; Davies & Laughton 1972; Roberts 1975).

Characteristics of sediment ridges include marginal moats adjacent to local highs, a distinctive wave-like surface (up to 2 km wavelength), ripples (down to 10 cm wavelength) and furrows (1–100 m in width). The detailed nature of these features has been studied by coring and by near-bottom towed vehicles housing sonar, photography and a high-resolution seismic profiler (Hollister, Southard, Flood & Lonsdale 1976). The waves and ripples apparently reflect constructional or mobile bedforms either parallel or transverse to the main current direction. In contrast, the furrows appear to result from erosion by helical vortices of well mixed bottom water orientated along the flow (Hollister, Southard, Flood & Lonsdale 1976).

It should be noted that the differential deposition by ocean bottom currents is pervasive and its characteristic features may be subdued by, or superimposed upon, the other geological processes that mould the slope and rise.

THE MARGIN AROUND THE BRITISH ISLES

The continental margin around the British Isles (Roberts, Hunter & Laughton 1977) includes the broad epicontinental shelves of the North Sea, Irish Sea, English Channel and Celtic Sea, the narrower Irish, Malin, Hebrides and West Shetland shelves, partly separated from the Irish Shelf by Porcupine Seabight, the 2000 m deep Rockall Trough and the associated 1000 m deep Faeroe–Shetland Channel, the almost submerged Rockall Plateau and the insular shelf of the Faeroes. To the northwest, the Iceland and Norwegian Basins, underlain by oceanic crust, are separated by the oceanographically important sill of the Iceland-Faeroe Rise which was created by the excessive magma production from the Iceland hotspot.

In this region of the northeast Atlantic, the presence of these anomalously shallow plateaux, separated by troughs and steep gradients precludes a simple description in terms of slope and rise provinces. Indeed, the distribution and major relief of these features are primarily due to the complex structural evolution of the area and their present morphology and depth to subsequent sedimentation and subsidence.

The main topographic units reflect the presence of both continental fragments and oceanic crust in the region (figure 3). The foundered Rockall Plateau, and at least part of the basement underlying the Faeroes, are fragments of continental crust. Their present distribution is a direct consequence of three distinct phases in the opening of the North Atlantic Ocean (Laughton 1975). During the first phase, in Lower Cretaceous time, the Rockall Trough and Faeroe–Shetland Channel opened, spreading the Greenland-Rockall–North American block away from Europe. In the second phase, at about 76 Ma B.P., the Labrador Sea opened, spreading the Greenland–Rockall block away from North America. In the final phase, beginning at about 60 Ma B.P., the Iceland Basin and Norwegian Basin were opened, so separating Greenland and the Rockall Plateau. The present distribution of the major relief of the margin was largely shaped by 60 Ma B.P., though the Iceland Basin has continued to widen since then.

The present morphology and physiography largely reflects the differential deposition of pelagic sediments by ocean bottom currents. The distribution of these sediments (figure 4) bears a well known and close relationship to the deep circulation of the North Atlantic Ocean (Jones et al. 1970; Davies & Laughton 1972; Roberts 1975). At the present, saline surface-water flows northward into the Norwegian Sea where it cools and becomes denser. This water sinks and ultimately flows back intermittently into the North Atlantic Basins over the Iceland–Faeroes Ridge, the Wyville–Thomson Ridge and through the Faeroe Bank Channel. In the Iceland Basin the overflow water, with sediment entrained in suspension, flows parallel to the continental slope south of Iceland and then parallel to the Reykjanes Ridge. Sediments deposited from this current have smoothed the irregular basement relief of the Iceland Basin and constructed the large Gardar Ridge sediment drift (Johnson & Schneider 1969; Ruddiman 1972). In the Rockall Trough, intermittent overflow across the Wyville–Thomson Ridge flowing southward along the west side of the Rockall Trough has contructed the Feni Ridge sediment drift (Ellett & Roberts 1973). The Feni Ridge has many of the typical features of sediment drifts including giant waves and possibly furrows. South of the Rockall Plateau the drift divides into a southward trending spur and another ridge that follows the contours of the Plateau to the northwest. Sediment drifts are also responsible for the physiography of much of the Rockall Plateau and have also been identified on the slope of North Biscay (Auffret, Pastouret & Kerbrat 1975).

In contrast to these areas, the physiography of the continental slope to the north, west and southwest of the British Isles has been closely controlled by the volume of sediments transported outward across the shelf and from the land areas. North of the North Sea, the continental slope is very gentle and underlain by thick sediments and there are no canyons. However, in the Faeroe–Shetland Channel strong currents may have eroded the sediments or prevented deposition. The slope west of Scotland and Ireland is steep with little sediment cover and only very few canyons. The sediment transport paths on the adjacent shelf lie mainly parallel to the shelf edge so that little sediment may be carried across the margin into the Rockall Trough (Kenyon & Stride 1971). However, in the eastern part of the Trough, fans comprised of terrigenous sediments are present and may have been supplied by sediments transported across the shelf between the Outer Hebrides and Ireland (Roberts 1975).

In contrast to the more northerly margins, the margin southwest of the British Isles is deeply incised by many canyons. Sediments on the adjacent shelf are transported towards the shelf edge and into the heads of canyons. Erosion is active in the canyons, and the adjacent rise and abyssal plain of the Bay of Biscay are floored by sediments deposited from turbidity currents. Examination of seismic reflexion profiles and long-range sonographs across and parallel to the margin has shown that the tectonic fabric associated with the initial rifting has influenced the course of many canyons.

Conclusion

Although many of the processes responsible for shaping the continental margins of the world are known in general terms, details are far from clear. In the progression of ocean engineering into deeper water, precise information will be needed on the nature and shape of the bottom, the geotechnical properties, the processes which might alter it during the lifetime of a structure and on the currents in the water above the bottom. These may vary significantly over short distances and detailed surveys will be necessary of both the morphology and the recent geological history. To give the required detail, near-bottom survey techniques will be necessary as well as narrow beam and precision surface measurements.

In this review, reference could not be made to all the papers written on continental margin morphology and its evolution. But four important books should be mentioned which could provide the reader with either reviews of the field or important collections of papers recently published following symposia (Heezen & Hollister 1971; Burk & Drake 1974; Woodland 1975; Vanney 1977).

References (Laughton & Roberts)

Auffret, G. A., Pastouret, L. & Kerbrat, R. 1975 *Abstr. 9th Int. Congress Sedimentologie, Nice 1975*, Theme 6, 1.
Belderson, R. H. & Kenyon, N. H. 1976 *Mar. Geol.* **22**, M69–M74.
Belderson, R. H., Kenyon, N. H. & Wilson, J. B. 1973 *Palaeogeogr. Palaeoclimatol. Palaeoecol.* **13**, 215–224.
Boillot, G., Dupeuple, P. A., Hennequin-Marchand, I., Lamboy, M., Lepretre, J.-P. & Musellec, P. 1974 *Rev. Géogr. phys. Géol. dyn.* **16**, 75–86.
Burk, C. A. & Drake, C. L. (eds) 1974 *The geology of continental margins.* New York: Springer-Verlag, 1009 pages.
Coulter, H. W. & Migliaccio, R. R. 1966 *U.S. geol. Surv. prof. Pap.* 542-C, 35 pages.
Curray, J. R. & Moore, D. G. 1971 *Bull. geol. Soc. Am.* **82**, 563–572.
Davies, T. A. & Laughton, A. S. 1972 In *Initial reports of the Deep Sea Drilling Project* (A. S. Laughton, W. A. Berggren *et al.*), vol. 12, pp. 905–934. Washington, D.C.: U.S. Government Printing Office.
Ellett, D. J. & Roberts, D. G. 1973 *Deep Sea Res.* **20**, 819–835.
Emery, K. O. 1971 In *The geology of the east Atlantic continental margin*, I.C.S.U./S.C.O.R. Working Party 31 Symposium, Cambridge, 1970 (ed. F. M. Delaney), vol. 1 (General and economic papers), pp. 3–29.

Emery, K. O., Niino, H. & Sullivan, B. 1971 In *The late Cenozoic glacial ages* (ed. K. K. Turekian), vol. 14, pp. 381–390. Yale University Press.

Falvey, D. A. 1972 *Aust. Petr. exp. Ass. J.* **12**, 86–88.

Flemming, N. C. & Roberts, D. G. 1973 *Nature, Lond.* **243**, 19–22.

Heezen, B. C. 1956 *Boln Soc. geogr. Colombia* **51/52**, 135–143.

Heezen, B. C. & Drake, C. L. 1964 *Bull. Am. Ass. Petrol. Geol.* **48**, 221–233.

Heezen, B. C., Ericson, D. B. & Ewing, M. 1954 *Deep Sea Res.* **1**, 193–202.

Heezen, B. C. & Ewing, M. 1952 *Am. J. Sci.* **250**, 849–873.

Heezen, B. C. & Ewing, M. 1955 *Bull. Am. Ass. Petrol. Geol.* **39**, 2505–2514.

Heezen, B. C. & Hollister, C. D. 1971 *The face of the deep.* London: Oxford University Press, 659 pages.

Heezen, B. C., Hollister, C. D. & Ruddiman, W. F. 1966 *Science, N.Y.* **152**, 502–508.

Heezen, B. C., Johnson, G. L. & Hollister, C. D. 1969 *Can. J. Earth Sci.* **6**, 1441–1453.

Heezen, B. C., Tharp, M. & Ewing, M. 1959 *Geol. Soc. Am. spec. Pap.* **65**, 122 pages.

Holtedahl, H. & Sellevoll, M. A. 1971 In *The geology of the east Atlantic continental margin*, I.C.S.U./S.C.O.R. Working Party 31 Symposium, Cambridge 1970 (ed. F. M. Delaney), vol. 2 (Europe), pp. 32–52. Inst. Geol. Sci. rep. no. 70/14.

Holtedahl, H. & Sellevoll, M. 1972 *R. Swed. Acad. Sci. Ambio spec. Rep.* **2**, 31–18.

Hollister, C. D. & Heezen, B. C. 1972 In *Studies in physical oceanography* (ed. A. L. Gordon), vol. 2, pp. 37–66. New York: Gordon and Breach.

Hollister, C. D., Southard, J. B., Flood, R. D. & Lonsdale, P. F. 1976 In *The benthic boundary layer* (ed. I. N. McCave), pp. 183–204. New York: Plenum.

Johnson, G. L. & Schneider, E. D. 1969 *Earth planet. Sci. Lett.* **6**, 416–422.

Jones, E. J. W., Ewing, M., Ewing, J. I. & Eittreim, S. L. 1970 *J. geophys. Res.* **75**, 1655–1680.

Kenyon, N. H. & Stride, A. H. 1970 *Sedimentology* **14**, 159–173.

Laughton, A. S. 1975 *Norg. geol. Unders.* **316**, 169–193.

Laughton, A. S., Berggren, W. A. *et al.* 1972 *Initial reports of the Deep Sea Drilling Project*, vol. 12. Washington, D.C.: U.S. Government Printing Office, 1243 pages.

Le Pichon, X., Francheteau, J. & Bonnin, J. 1973 *Developments in geotectonics*, vol. 6. Amsterdam: Elsevier, 300 pages.

Menard, H. W. & Smith, S. M. 1966 *J. geophys. Res.* **71**, 4305–4325.

Montadert, L., Roberts, D. G. *et al.* 1977 *Nature, Lond.* **268**, 305–309.

Roberts, D. G. 1972 *Mar. Geol.* **13**, 225–237.

Roberts, D. G. 1975 *Phil. Trans. R. Soc. Lond.* A **278**, 447–509.

Roberts, D. G. & Caston, V. N. D. 1975 *Proc. 9th World Petroleum Congress, Tokyo*, vol. 2, pp. 281–288. London: Applied Science Publishers.

Roberts, D. G., Hunter, P. M. & Laughton, A. S. 1977 Sheet 2, *Continental margin around the British Isles*, chart C 6567. Published by Hydrographic Department, U.K.

Ruddiman, W. F. 1972 *Bull. geol. Soc. Am.* **83**, 2039–2062.

Shepard, F. P. 1972 *Earth Sci. Rev.* **8**, 1–12.

Shepard, F. P. 1975 *Mar. Geol.* **19**, 131–138.

Shepard, F. P. 1976 *J. Geol.* **84**, 343–350.

Shepard, F. P. & Dill, R. F. 1966 *Submarine canyons and other sea valleys.* Chicago: Rand McNally, 381 pages.

Shepard, F. P., Marshall, N. F. & McLoughlin, P. A. 1974 *Deep Sea Res.* **21**, 691–706.

Vanney, J.-R. 1977 *Géomorphologie des plates-formes continentales.* Paris: Doin editeurs, 300 pages.

Woodland, A. W. 1975 *Petroleum and the continental shelf of northwest Europe.* Romford, Essex: Applied Science Publishers, 501 pages.

Phil. Trans. R. Soc. Lond. A. **290**, 87–98 (1978)   [87]

Printed in Great Britain

Currents on continental margins and beyond

By W. J. Gould

Institute of Oceanographic Sciences, Wormley, Godalming, Surrey GU8 5UB, U.K.

A description is given of the techniques at present in general use for the measurement of currents with particular emphasis on methods used in water depths greater than 200 m. The general characteristics of current motions both in the deep ocean and in continental shelf seas are described and are categorized in terms of their energetics, periodicities and vertical and horizontal length scales.

INTRODUCTION

The continental slope is a boundary region between the tidally dominated current régime on the shelf and the generally less energetic but more complex currents in the deep ocean. Currents in this transition region are difficult to characterize since the local variations in water depth are great and thus have a large influence on current patterns and energetics. In this paper we shall consider the characteristics of currents in the deep ocean and on the continental slope, and where relevant compare these with the currents in the shallow continental shelf seas. Apart from a few intensive studies of deep ocean currents, rather little is known of their regional variability, a result largely of the immense horizontal extent of the abyssal ocean and the high cost and sophistication of the instrumentation needed to make measurements in the deep ocean.

Much is now known of the temporal variability of ocean currents and of their vertical structure, and a large part of this paper will be devoted to the types of motion present over a range of periodicities from months to seconds.

CURRENT MEASURING TECHNIQUES

In order to understand the problems of making measurements of currents in the ocean it will be informative to consider the techniques now in general use for making such measurements, the history of their development and, above all, their shortcomings.

Eulerian measurements

This class of measurements includes all measurements of velocity–time-series at fixed points in the ocean. Before the mid-1960s there were almost no measurements of currents away from the continental shelf (Bowden 1954). This situation was changed by the development of self-contained, internally recording current meters which could make measurements at any depth for periods of weeks and months (Richardson, Stimson & Wilkins 1963; Aanderaa 1964). These and other similar instruments employed propellers or rotors to measure current speed and a magnetic compass and vane to determine flow direction. The instruments rapidly achieved a high degree of reliability which was unfortunately not matched by the reliability of the mooring systems on which they were deployed. In the deep ocean, moorings capable of withstanding months of exposure have become routine only over the past 5 years or so. The engineering problems encountered by one U.S. laboratory are well documented by Heinmiller (1976).

The types of mooring now in general use have evolved with the development of new materials and with the understanding of the effects of environmental conditions on the performance of instruments. In the past **3** or **4** years it has been shown that under certain circumstances the records of currents made with conventional rotor/vane instruments on moorings with buoyancy at the sea surface may significantly overestimate the magnitude of the currents. The problem is due to energy at surface-wave frequencies being transmitted down the mooring line and producing motions of a frequency too high for the instrument to respond adequately. Paradoxically

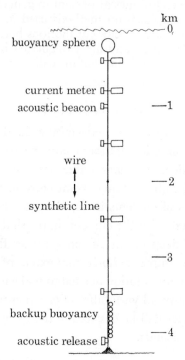

FIGURE 1. Design of current meter mooring for measurements in deep water. Typical main buoyancy would be around 700 kg and anchor mass 1000 kg.

the effects are most marked at the deeper levels, due to the generally low current speeds at those depths, and may not be so severe in the energetic, low-frequency near-surface flows. A comparison of current statistics at the same site from both surface and subsurface moorings (Gould & Sambuco 1975) showed that current amplitudes could be overestimated by as much as 100 % over a depth range from 100 to 2000 m in water of 2600 m depth. As a general precaution the use of conventional rotor/vane current meters on surface buoy moorings should be avoided.

Where measurements are not required in the uppermost layers of the ocean a design of subsurface mooring similar to that shown in figure 1 has generally been adopted. The mooring is recovered via an acoustic release between the instrumented part of the mooring line and the anchor. The upper 1500 m is made up of 6 and 8 mm diameter torque balanced wire terminated with swaged steel fittings. The wire is galvanized and may also be jacketed by a plastic coating. In very deep water (greater than 2000 m) the weight of an all-wire mooring is often prohibitive and in this case the lower sections of mooring line may be made up of braided man-made fibres. This is a compromise since the large cross-sectional area of the 'synthetic' line leads to high drag and the poorly defined stretch characteristics may mean that instruments cannot be placed at precisely the correct depth.

The aim is to make the mooring as 'stiff' as possible without reducing the safety factor of the mooring line. A 'stiff' mooring will remain upright and not be displaced significantly from the vertical by the currents which pass it. A typical I.O.S. deep-water mooring in 5000 m of water experiences vertical excursions at its upper end of the order of ± 10 m at tidal frequencies and is subject to stretch of the order of 10 m over an exposure period of four months.

A solution to many of the problems of mooring design may lie in the use of the new aramid fibre materials (marketed under the trade name Kevlar). These have similar densities to other man-made fibres but have strengths and stretch characteristics similar to steel. Present drawbacks to such an approach lie in the high cost and the need to armour such materials in the upper ocean owing to the possibility of damage caused by fish bite (Stimson 1965; Haedrich 1965).

In the majority of long exposure moorings in the deep sea, buoyancy is inserted in the mooring line immediately above the acoustic release. In the event of a mooring failure in the upper part of the line, sufficient buoyancy remains to bring the instruments back to the sea surface and enable the cause of failure to be investigated.

Moorings of the type described above have been used extensively in water depths between 200 and 5000 m. The problems of making current measurements in the uppermost, wave-affected depth range and from moorings with surface buoyancy will most probably be overcome by the introduction of electromagnetic and acoustic velocity sensors which can measure directly the orthogonal components of current velocity (Tucker 1972).

A rather different set of mooring design problems has led to another technique for deploying current meters in the shelf seas. Although the instruments are basically the same as those used in deep water, they are not in general used on single-point moorings. Poor acoustic conditions make the use of subsurface buoyancy coupled with acoustic releases difficult and instead a U-shaped design is employed. This consists of an instrumented leg with subsurface buoyancy connected by a ground line to a surface marker buoy (Howarth 1975 a). The relatively shallow water depths make the use of large diameter wires possible and there are thus no problems in lifting the full anchor weight when recovery is undertaken.

The major hazard to such installations comes from fishing activity. In the event of loss of the surface marker buoy the instruments are recovered by dragging for the ground line. Largely as a result of fishing activity the typical deployment periods of shallow water moorings in U.K. waters are of the order of 2 months or less.

Lagrangian techniques

Into this category fall all the methods that involve the tracking of a drogue or float intended to tag a body of water. At the same time that recording current meters were being developed, the first measurements of currents in the deep ocean were being made by tracking neutrally buoyant floats (Swallow 1955). These devices are designed to be less compressible than seawater. They are ballasted to sink at the sea surface but as they sink they gain buoyancy until at some predetermined pressure level the density of the float is equal to the density of the surrounding water. They remain in stable equilibrium at this level and are tracked by their emission of acoustic signals. In the earlier floats these signals were emitted continuously at a regular repetition rate and the float position was determined by manoeuvring the ship overhead of the float. This seriously limited the number of floats that could be tracked at any time and the area over which measurement could be made.

The development and use of a transponding float, in which float ranges from the attendant ship or from a fixed interrogator on the seabed, made possible the tracking of floats over distances of up to 70 km (Swallow, McCartney & Millard 1974).

As with all ship-based tracking systems the duration of the measurement is limited by the endurance of the attendant ship (typically 30 days or less). Using low-frequency sound acoustic ranges of many hundreds of kilometres are attainable and lead to the possibility of tracking neutrally buoyant floats from fixed shorebased or moored listening stations (Rossby & Webb 1970). The sound signals are channelled by a minimum in the vertical profile of sound velocity at a depth generally near 1000 m (the Sofar channel).

Tracks of many months duration from a large area of the western North Atlantic SW of Bermuda have been collected since 1973. For the most part these have been from depths close to the Sofar axis but it now appears that reception conditions may be adequate over a range of depths from 500 to 3000 m.

LENGTH AND TIME SCALES OF CURRENT VARIABILITY

Both in shelf seas and in the deep ocean a measurement of current samples a range of periodicities spanning from a few seconds to hundreds of days. Some of the frequency components are discrete such as tidal motions, the majority are not so and fall in a range of frequencies. Similarly the vertical structure of the current profile and the horizontal scales of current variability are dependent on the dominant motions present and on factors such as the stratification of the water column and the local variation of water depth.

We shall attempt to characterize the various types of motion known to exist in the ocean, to identify their causes and to identify their dominant time scales and vertical and horizontal length scales.

Table 1 gives a summary of these characteristics for the majority of known current motions. The figures quoted for periods, scales and amplitudes are intended only to give orders of magnitude. There is, for instance, no lower limit on the amplitude of signals but in most cases typical values have been quoted.

The mean circulation

Undoubtedly if a very long term average of the circulation of the ocean were obtainable a mean value would emerge. In much of the ocean the magnitude of the mean flow would be small. The strongest and best-defined mean flows in the ocean are found in the western boundary currents (e.g. the Gulf Stream and the Kuroshio) and in the deep flows between ocean basins (e.g. the outflow from the Norwegian Sea into the North Atlantic Ocean and the in and out flows between the Mediterranean Sea and the Atlantic Ocean).

Even these well-defined flows, although they may be energetic, are far from steady. The Gulf Stream varies its position and strength from day to day (Robinson 1971) and similar variability is found in other western boundary currents.

The direct measurement of mean flows in the deep ocean far removed from major currents has only recently been possible with the accumulation of very long direct current measurements. Schmitz (1976a, b, 1977) has computed the mean values from current meter records in the deep western North Atlantic. The records are several hundred days in length and from them it appears that estimates of the mean current begin to stabilize only after an averaging period of the order of 200–300 days.

In continental shelf seas the lowest-frequency circulation patterns are constrained by topography and affected by seasonal changes much more than is likely in the deep ocean. It is unlikely that the record-duration needed to define the mean is appreciably different from that found by Schmitz. There are additional problems in defining the mean circulation of shelf seas owing to the presence of the very strong tidal oscillations.

TABLE 1

motion	period	horizontal wavelength	vertical wavelength	typical amplitude cm/s	geographical area
mean flow	—	various	h†	0–10	all seas and oceans
climatic change	10 a	global	h	?	all seas and oceans
seasonal variability	1 a	global	h	?	all seas and oceans
mesoscale activity	30–100 d	100–500 km	h	5–100	all deep oceans
meteorological disturbances	aperiodic	basin wide in shelf seas	h	up to 100	shelf seas and upper layer of ocean
shelf and edge waves	2–10 d	10 km across slope 500 km along slope	h	5–10	on topographic features, e.g. seamounts, shelf edges, etc.
inertial oscillations	12 h/sin L† (0.5–5 d)	tens of kilometres	100 m	5–50	all seas and oceans
tidal oscillations	0.5–1 d	basin width	h	10–20, deep sea 10–200, shelf seas	all seas and oceans
internal waves	inertial period to 10 min	1 km	100 m	5	throughout the stratified ocean
surface waves	1–20 s	100 m	limited to upper 30 m of water column	up to 200	all seas and oceans

† L = latitude; h = water depth; a = year.

The overflows from deep channels between ocean basins may retain their identity far from the strait from which they issue. No direct measurements have been made which support the existence of such flow but they may be inferred from water mass properties and from sediment distributions. Roberts, Hogg, Bishop & Flewellen (1974), in investigating sediment ridges in the Rockall area, infer the presence of a mean southward flow of water in that area originating from the Norwegian Sea outflow. The flows may well retain some of the energetic characteristics of the contained channel flows.

Climatic change and seasonal variability

The very long period changes associated with the changes in world climate are not, as yet, identifiable in the relatively short records of oceanographic variables that are available and certainly not in the direct measurements of current. Large changes, with periods longer than a year but shorter than the periods of climatic change, are known. The best documented is that of the El Niño phenomenon off the coast of Peru which is connected with long-period, atmospherically coupled changes in the equatorial currents (Wyrtki 1975).

Seasonal variations in the ocean are most readily observable in the upper ocean thermal structure and in the effect of runoff from the land on surface salinities. It is not unreasonable to suppose that these factors also affect circulation patterns in a seasonal way but the effects are for the most part small. A very striking example, however, is the effect of the seasonal monsoon winds of the Indian Ocean on the Somali current off the coast of east Africa. The Somali current (the Indian Ocean analogue of the Gulf Stream) reverses direction seasonally in response to the monsoon winds (Leetmaa & Truesdale 1972; Leetmaa 1973). This and the El Niño are, however, unique events and seasonal and climatic fluctuation are in general not nearly so dramatic.

Mesoscale variability

The earliest observations of deep currents in the ocean with the use of neutrally buoyant floats were expected to reveal the slow mean circulation pattern. The measurements, however, showed that velocities of many centimetres per second were common and, above all, that the flows were not unidirectional over large distances but rather showed a dominant spatial scale of several tens of kilometres (Crease 1962).

There was a gap of many years before full-scale experiments were mounted to investigate further these energetic mesoscale motions. The Russians conducted a series of Polygon experiments in various seas and oceans (Fofonoff 1976) culminating in a seven-month experiment in the North Atlantic in 1970 with 17 current-meter moorings in a 2° square (Brekhovskikh *et al.* 1971). This was followed in 1973 by the joint U.S.A./U.K. Mid-Ocean Dynamics Experiment (MODE-1). This latter experiment used both moored current meters and ship- and Sofar-tracked neutrally buoyant floats (MODE Scientific Council 1973; Wunsch 1976), together with a wide variety of other oceanographic instruments (Gould 1976). (The main results of the MODE-1 experiment are presented in Simmons *et al.* (in preparation).)

The observations in both the Polygon and MODE-1 experiments revealed features in the low-frequency circulation patterns with velocities as high as 50 cm/s in the upper ocean and with horizontal dimensions between 100 and 140 km. The term 'eddies' has been generally applied to such features and is in common use; however, the implication that the features are closed and eddy-like is not necessarily valid. In the deeper layers the eddies are found to have smaller horizontal dimensions and to be generally less energetic.

It is often difficult to visualize the structure of deep ocean eddies from the discrete and often randomly distributed current velocity vectors from both neutrally buoyant floats and current meters. A technique known as 'objective analysis' has been used by Freeland & Gould (1976) to produce a stream function map from these discrete vectors. Two examples of these maps are shown in figure 2. They represent the flow fields at 500 and 1500 m for a 3-day period during the MODE-1 experiment. The flow fields at the two levels are clearly different and an analysis by Freeland, Rhines & Rossby (1975) shows that the circulation pattern at 500 m and at the deeper levels progress westwards at different speeds, the lower levels having the higher phase velocities.

It is now becoming clear that a range of eddy-like features of different sizes and energetics exist in different parts of the deep ocean. The smallest and most energetic are the cyclonic and anticyclonic rings formed by the 'pinching off' of meanders in the Gulf Stream (Richardson, Cheney & Mantini 1977); some of these may penetrate into the interior of the ocean (Parker 1971) but their distribution seems to be restricted to certain well-defined areas. Larger but less energetic features have been identified in the interior of the Atlantic Ocean, for the most part by the isotherm displacements seen in hydrographic or expendable bathythermograph (x.b.t.) sections.

Data from x.b.ts have been used by Dantzler (1977) to identify the most energetic eddy regions of the North Atlantic. The results show large (factor of 4) changes in eddy potential energies over very small horizontal distances in some parts of the ocean. It is possible that these could be associated with major topographic features.

It is to be expected that in situations where energetic mid-ocean mesoscale features impinge upon mid-ocean ridges, sea mounts or continental slopes, the reduction in water depth would lead to an enhancement of velocities. An example of this has been documented by Freeland & Dow (1976) who monitored the track of Sofar floats as they executed a series of violent oscillations over the Blake Bahama outer rise. The floats made roughly circular tracks 50 km in diameter with periods of between 17 and $3\frac{1}{2}$ days (speeds of 12–30 cm/s), before following the depth contours to the south at speeds as high as 45 cm/s. The floats were at 2000 m depth.

It is unlikely that mesoscale eddies could penetrate onto the continental shelf without their energy being largely dissipated but there are examples of Gulf Stream rings penetrating well up the continental slope off the eastern seaboard of the United States (Bisagni 1976).

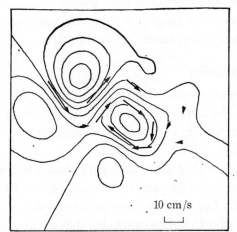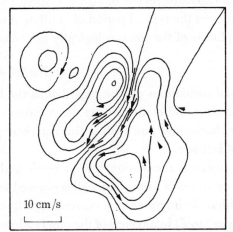

FIGURE 2. Streamfunction maps for a 3-day period of the MODE-1 experiment derived from current observations shown by arrows. Each box is 600 × 600 km. Data are from 500 m (left) with contour interval of 200 cm km s^{-1} and 1500 m (right) with contour interval of 50 cm km s^{-1}. Source data are from both current meters and neutrally buoyant floats, low-pass filtered and averaged over 3 days.

Meteorological disturbances

The passage of atmospheric depressions across areas of continental shelf can, by the action of pressure and wind stress on the sea surface, give rise to anomalously high (and low) water levels (storm surges). Associated with the elevation of the sea surface there are often large perturbations of the generally tidally dominated flow patterns. The mechanism of surge generation is quite well understood and from forecast meteorological conditions computer models can be used to predict both surge heights and current strengths. The validity of the model may be assessed by hindcasting the observed surge heights at coastal stations (Flather & Davies 1976).

There seem to be very few well-documented examples of directly measured current disturbances associated with meteorological forcing. Howarth (1975 b) presents current data from the southern Irish Sea in which a departure of over 50 cm/s from the normal tidal flow is attributed to meteorological forcing. Caston (1976) documents currents measured in the southern North Sea and shows the correlation between flow and surface winds. The strong effect of local bottom topography is evident in this data.

R. A. Flather (personal communication) has compared measured currents in the North Sea during a severe storm in January 1976 with the predicted surge currents from his computer model. Good agreement has been found. It is clear that such meteorologically induced flows are of great importance in shelf seas but the number of cases in which adequate current measurements are available to test computer models is very small.

In addition to the direct forcing the passage of meteorological disturbances may initiate resonant motions trapped along depth discontinuities. The continental shelf edge is an obvious area in which these effects could be found and a considerable number of workers, particularly those involved in the field of coastal upwelling, have studied this phenomenon. Once again the number of examples of direct current measurements in which these oscillations are detectable is small; for the most part the trapped waves are seen in the spectra of sea levels and are forced by changes in the surface wind stress. Kundu, Allen & Smith (1975), Huyer et al. (1975) and Kundu & Allen (1976) have all investigated this phenomenon along the coasts of Oregon and Washington State where additional current measurements are available. Clarke (1977) has made an extensive review of all the evidence for the existence of edge waves.

In all cases the typical period of oscillation was in the range 3–6 days and, where measured, the amplitude of the meteorologically forced current was small (ca. 10–15 cm/s).

Inertial oscillations

Inertial oscillations can be regarded as the free response of the ocean to an impulsive force at the surface and controlled by the rotation of the Earth. The motions are for the most part circular, horizontal and with a characteristic period (in hours) $T = 12/\sin \lambda$, where λ is the angle of latitude.

In many cases the motions are energetic and since they have, at any position, a well-defined frequency which is in general well separated from tidal frequencies they have proved a subject of considerable interest when current records of a few days duration are available.

Webster (1968) in a review of the observations up to that date concluded that the phenomenon was transient and of very thin vertical extent and remarked that the generation mechanism was not probably related to the passage of storms. A considerable advance was made by the work of Pollard (1970) and Pollard & Millard (1970). In these papers a simple model of inertial current generation by a changing wind stress is used to hindcast the measured inertial oscillations on two buoys at which wind observations were also available.

The model predictions are in good general agreement with the observations. The intermittent nature of the inertial signal is reproduced with energetic oscillations persisting for between 2 and 5 days before decaying.

The model does not account for the relatively high level of inertial activity found at depth. Pollard analysed records of 200 days duration from depths between 10 and 2000 m in 2600 m. He found amplitudes of inertial oscillations were approximately uniform over a depth range of 50–500 m, that at 10 m the amplitude was approximately twice as large, and at 1000 and 2000 m approximately half as large. Amplitudes at the deepest two levels were 10 cm/s maximum and for 10 % of the time the amplitude exceeded 3–4 cm/s.

Perkins (1972) analysed records of approximately 50 days duration from a site in the western Mediterranean. Five records were obtained at levels every 500 m from 200 to 2200 m.

Persistent oscillations at a frequency 3 % above the local inertial value were found at all levels. Almost all the energy had a clockwise polarization. Energies varied very little over the deepest levels but were much less (factor of 4) than at 200 m. This simple deep structure was attributed to the very weak stratification below about 500 m.

Recent techniques of current profiling (Sanford 1971) have led to more detailed studies of the vertical structure of inertial oscillations (Leaman & Sanford 1975). Vertical profiles obtained one-half an inertial period apart are averaged and then subtracted from the original profiles. This leads to the elimination of lower wavenumber signals and enables a study of the wavenumber spectrum of the inertial oscillations to be made. The calculations have been done in terms of 'stretched' coordinates since the local Brunt–Väisälä stability frequency has the effect of changing the vertical scale and also the amplitude of the signals.

Analysis of several profiles from the western North Atlantic shows again a dominance of clockwise polarization. The wavenumber spectrum has a maximum energy density at low wavenumbers corresponding to stretched wave lengths between 100 and 500 m and at higher wavenumbers a spectral slope of -2.5 on a log–log plot.

A further analysis of these data is presented by Leaman (1976). He concludes that the behaviour of inertial oscillations in the deep ocean is consistent with internal wave theory. The data show a downward flux of energy from the upper ocean at the inertial frequency. This flux persists into the deepest parts of the profiles (up to 5000 m) but its magnitude cannot be determined accurately owing to the uncertainty in measuring the exact frequency of the observed oscillations.

Inertial oscillations certainly exist in continental shelf seas but their occurrence is less well documented than for deep water. This may partly be due to the fact that shallow water records are in general dominated by tidal signals which may mask the inertial oscillations.

Currents of tidal period

The astronomical forcing of the attraction of the sun and moon induce surface elevations and currents in the ocean at well defined frequencies. These are almost entirely in the diurnal (period 24–25 h) and semidiurnal (period 12–12½ h) bands.

Energy at these tidal frequencies if found in virtually all current records and in shelf seas the tides represent by far the most energetic motions. In shelf seas the pattern of tidal currents is usually fairly simple and is controlled by the geometry of the adjacent land mass and the variations in bottom depth.

Perhaps the most detailed analysis of tidal currents at a fixed point in shallow water is that of Pugh & Vassie (1976). They have analysed year-long data of both tidal heights and current values from the Inner Dowsing Light tower. The current meter measurements are from approximately mid-depth in 20 m of water.

The analysis of the currents reveals that over 96 % of the variance is accounted for by the principal tidal constituents. The remaining 4 % is shared between the higher-frequency shallow-water tidal components and the residual aperiodic motions. The mean current over the total duration of the record has a magnitude of only 4 cm/s. No energy is seen in the current record at inertial frequencies.

In the Pugh & Vassie observations particularly, the currents behave in a regular and predictable way and are almost entirely deterministic. The currents in the deep ocean at tidal frequencies are quite different.

To date the most detailed analysis of tidal currents in the deep ocean is that by Hendry (1977). He has taken both current and temperature observations from the moored instruments in the MODE-1 experiment to investigate the vertical and horizontal structure of motions in the semidiurnal tidal frequency band.

Tidal currents in the deep ocean may be considered to have a component which is uniform and in phase over the entire water column (the barotropic component) and a component which varies with depth (baroclinic component). The baroclinic component can be resolved into motions with particular vertical modal structures having 1, 2, 3,... zero crossings (the first, second, third, etc., baroclinic modes). The depths of the zero crossings are a function of the vertical stratification of the water column.

Hendry analysed data from the MODE-1 field experiment to investigate the distribution of energy between the barotropic and baroclinic (internal) tide in the semidiurnal band of frequencies (periods between 12.00 and 12.86 h).

The tidal currents did not exceed an amplitude of *ca.* 5 cm/s at any depth and both the barotropic and baroclinic currents were found to be around 1 cm/s.

The baroclinic tides have short horizontal wavelengths and are generated by the interaction of the barotropic tide with the slopes of major features such as mid-ocean ridges and continental slopes. Their dependence on the stratification of the water column and their propagation into the interior of the ocean from a variety of source regions tends to make the baroclinic tide non-deterministic.

Internal waves

The internal tides are a particular case of internal waves. These can be shown to exist over a range of frequencies bounded at the lower end by the local inertial frequency and at the high frequency end by the Brunt–Väisälä frequency, N.

Values of N in the ocean vary between those corresponding to periods of 2–3 h in the deep ocean to a few minutes in the thermocline.

Freely propagating internal waves can only exist in a stratified ocean and within the specified range of frequencies. Although their vertical amplitude may reach values of tens of metres the associated horizontal current velocities, except perhaps in the tidal band, are generally no more than 1 or 2 cm/s.

A very detailed study of internal wave activity in the deep ocean was carried out in 1973 (Briscoe 1975) and involved the measurement of current velocity and temperatures along the legs of a tetrahedral, three-legged mooring in the Sargasso Sea. The horizontal kinetic energies at various depths in the internal wave band were found to be proportional to the Brunt–Väisälä frequency at the depth of measurement, i.e. the most energetic internal wave motions exist where the stratification of the water column is strongest.

The method of generation of internal wave activity in the ocean is most probably a combination of forcing induced at the sea surface and the interaction of currents with topographic features. Internal wave activity has characteristic signatures which can be recognized by the relation between particular spectral qualities. This ordered nature distinguishes the wave-like activity from random motions which exist in the ocean both in the internal wave band and at higher frequencies due to general turbulent motions. These are of such low amplitude as to be of no importance in most practical offshore activities.

Surface waves

In the uppermost layers of all oceans and seas, surface waves may provide the most energetic current component. The direct measurement of the wave orbital velocities in the open ocean poses considerable difficulties not only in the design of instruments which can adequately measure the high frequency motions but also in the design of stable platforms from which the measurements can be made.

The presence of these energetic, high-frequency motions in the upper layers has been found to have a serious effect on the measured values of current velocity at all depths on a surface following buoy when conventional rotor/vane current meters are used (Gould & Sambuco 1975). As a general rule, unless current meters with an adequate high-frequency response are available, the measurement of currents on moorings using a surface following buoy should be avoided. This problem means that there are few reliable measurements of currents in the uppermost, wave-affected layers of the ocean and results in a lack of adequate statistics of the currents likely to be encountered by surface piercing structures.

PARTICULAR PROBLEMS OF CONTINENTAL MARGINS

The range of ocean depths between 200 and 2000 m occupy the boundary between the distinctly different continental shelf and deep ocean current régimes and as such are subject to motions characteristic of both zones.

The continental slopes may be subject, for example, to both the energetic tidal motions originating on the shelf and the intermittent and unpredictable inertial and mesoscale motions from the deep sea. They may additionally, in some specific areas, be swept by deep overflowing currents from channels between ocean basins. The zone is thus one of energetic but largely unpredictable (except in a statistical sense) currents. The prospects of offshore engineering activity in the 200–2000 m depth zone will probably highlight the lack of sufficiently long current records from this zone of the ocean.

REFERENCES (Gould)

Aanderaa, I. 1964 *N.A.T.O. Subcommittee on Oceanographic Research, Techical Report*, no. 16, 46 pages.
Bisagni, J. 1976 *N.O.A.A. Dumpsite Evaluation Report* 76–1.
Bowden, K. F. 1954 *Deep Sea Res.* 2, 33–47.
Brekhovskikh, L. M., Fedorov, K. N., Fomin, L. M., Koshlyakov, M. N. & Yampolsky, A. D. 1971 *Deep Sea Res.* 18, 1189–1206.
Briscoe, M. G. 1975 *J. geophys. Res.* 80, 3872–3884.
Caston, V. N. D. *Estuar. coast. mar. Sci.* 4, 23–32.
Clarke, A. J. 1977 *J. phys. Oceanogr.* 7, 231–247.
Crease, J. 1962 *J. geophys. Res.* 67, 3175–3176.
Dantzler, H. L. 1977 *J. phys. Oceanogr.* 7, 512–519.
Flather, R. A. & Davies, A. M. 1976 *Q. Jl R. met. Soc.* 102, 123–132.
Fofonoff, N. 1976 *Oceanus* 19(3), 40–44.
Freeland, H. J., Rhines, P. B. & Rossby, T. 1975 *J. mar. Res.* 33, 383–404.
Freeland, H. & Dow, D. 1976 *Polymode News*, no. 1, 1 and 3 and figures. (Unpublished manuscript.)
Freeland, H. J. & Gould, W. J. 1976 *Deep Sea Res.* 23, 915–923.
Gould, W. J. & Sambuco, E. 1975 *Deep Sea Res.* 22, 55–62.
Gould, W. J. 1976 *Oceanus* 19(3), 54–64.
Haedrich, R. L. 1965 *Deep Sea Res.* 12, 773–776.

Heinmiller, R. H. 1976 *Woods Hole Oceanogr. Instn Tech. Rep.* 76–53, 73 pages.

Hendry, R. M. 1977 *Phil. Trans. R. Soc. Lond.* A **286**, 1–24.

Howarth, M. J. 1975 *a Technology of buoy mooring systems*. London: Society for Underwater Technology, 75–86.

Howarth, M. J. 1975 *b Estuar. coast. mar. Sci.* **3**, 57–70.

Huyer, A., Hickey, B. M., Smith, J. D., Smith, R. L. & Pillsbury, R. D. 1975 *J. geophys. Res.* **80**, 3495–3505.

Kundu, P. K., Allen, J. S. & Smith, R. L. 1975 *J. phys. Oceanogr.* **5**, 683–704.

Kundu, P. K. & Allen, J. S. 1976 *J. phys. Oceanogr.* **6**, 181–199.

Leaman, K. D. & Sanford, T. B. 1975 *J. geophys. Res.* **80**, 1975–1978.

Leaman, K. D. 1976 *J. phys. Oceanogr.* **6**, 894–908.

Leetmaa, A. & Truesdale, V. 1972 *J. geophys. Res.* **77**, 3281–3283.

Leetmaa, A. 1973 *Deep Sea Res.* **20**, 397–400.

MODE-1 Scientific Council 1973 MODE-1, the program and the plan. Department of Meteorology, Massachusetts Institute of Technology (unpublished manuscript).

Parker, C. E. 1971 *Deep Sea Res.* **18**, 981–993.

Perkins, H. 1972 *Deep Sea Res.* **19**, 289–296.

Pollard, R. T. 1970 *Deep Sea Res.* **17**, 795–812.

Pollard, R. T. & Millard, R. C. 1970 *Deep Sea Res.* **17**, 813–821.

Pugh, D. T. & Vassie, J. M. 1976 *Dt. hydrogr. Z.* **19**, 163–213.

Richardson, P. L., Cheney, R. E. & Mantini, L. A. 1977 *J. phys. Oceanogr.* **7**, 580–590.

Richardson, W. S., Stimson, P. B. & Wilkins, C. H. 1963 *Deep Sea Res.* **10**, 369–388.

Roberts, D. G., Hogg, N. G., Bishop, D. G. & Flewellen, C. G. 1974 *Deep Sea Res.* **21**, 175–184.

Robinson, A. R. 1971 *Phil. Trans. R. Soc. Lond.* A **270**, 351–370.

Rossby, T. & Webb, D. 1970 *Deep Sea Res.* **17**, 359–365.

Sanford, T. B. 1975 *J. geophys. Res.* **80**, 3861–3871.

Schmitz, W. J. 1976 *a J. geophys. Res.* **81**, 4981–4982.

Schmitz, W. J. 1976 *b Geophys. Res. Lett.* **3**, 373–374.

Schmitz, W. J. 1977 *J. mar. Res.* **35**, 21–28.

Stimson, P. B. 1965 *Deep Sea Res.* **12**, 1–8.

Swallow, J. C. 1955 *Deep Sea Res.* **3**, 74–81.

Swallow, J. C., McCartney, B. S. & Millard, N. W. 1974 *Deep Sea Res.* **21**, 573–595.

Tucker, M. J. 1972 *Proc. Soc. underwat. Technol.* **2**, 53–58.

Webster, F. 1968 *Rev. Geophys.* **6**, 473–490.

Wunsch, C. 1976 *Oceanus* **19**(3), 45–53.

Wyrtki, K. 1975 *J. phys. Oceanogr.* **5**, 572–584.

Phil. Trans. R. Soc. Lond. A. **290**, 99–111 (1978)   [99]

Printed in Great Britain

Offshore subsea engineering

By E. C. Goldman

Exploration and Production Department, Shell Internationale Petroleum
Maatschappij B.V., The Hague, The Netherlands

As a result of years of research and development work on subsea completions, manifolds and flowlines, followed by field installations in relatively shallow and calm waters, a number of well completion and flowline laying and connecting methods are available and operational. The installation methods are basically diverless and control and maintenance are achieved by sophisticated systems; likewise the floating offshore terminals such as the E.L.S.B.M. and the Spar have been designed, built and put into operation.

Although these systems were meant to be installed mainly in deeper waters, say beyond 210 m where diver access and the use of conventional techniques would become rather limited for technical and economic reasons, it is now evident that many fields in shallower waters can be more economically and efficiently developed or complemented by the use of deepwater techniques.

Current development work concentrates on the design and evaluation of multibore production risers, floating production platforms and pipelaying and repair.

My subject takes us to one of the furthest frontiers yet reached by modern oil technology. Before I discuss in a little detail the latest trends I should, I think, make the general point that although advances, stimulated particularly by the challenge of the North Sea, have been very rapid in recent years, the industry's capacity to produce oil and gas in deep water is still in its infancy.

As figure 1 shows, it is now possible to drill exploration wells from semi-submersible drilling rigs or drill-ships in more than 1000 m of water. This takes us right off the continental shelf upon which offshore operations have hitherto been concentrated and on to the steeply falling continental slope. A Shell company has drilled a well in 720 m of water off West Africa; Exxon have more recently gone deeper still – 1000 m offshore Thailand. These efforts, however, have so far not met any commercial success.

By contrast, for production drilling the present limit is about 200 m, the depth at which the Brent and Statfjord fields northeast of the Shetland Islands are being developed. There we still have to create an oasis of dry land above the sea as a base on which the work can be done. This oasis has been provided by huge platforms of steel and concrete upon which all the facilities for oil, gas and water production and separation can be established. These units are tremendously costly, particularly in the current climate of inflation. To take one example: since 1972 when the development of Brent began, the investment per daily barrel of peak producing capacity has doubled, to between \$6000 and \$9000 for the new generation of North Sea fields.

The four Brent production platforms and their facilities alone will together cost more than \$10^9$. For Statfjord, in the Norwegian sector, the biggest North Sea field yet found, the cost of the initial platform will be more staggering still.

It is well known that from each of these platforms it will be possible to drill some 40 deviated wells – that is to say wells which at depth will be widely spaced around the platform to tap the oil reservoir at a number of points. The wells are drilled vertically at first and then at an angle, to reach their required targets in the reservoir. The wells reaching out furthest will require a maximum angle of about 65 degrees and be 3 km from the platform. Even deviated wells are unable, however, to reach the whole of a very large reservoir, or isolated accumulations that are too small to justify individual platforms. So the basic problem is how to find means of completing and producing remote wells at costs that will meet our economic criteria.

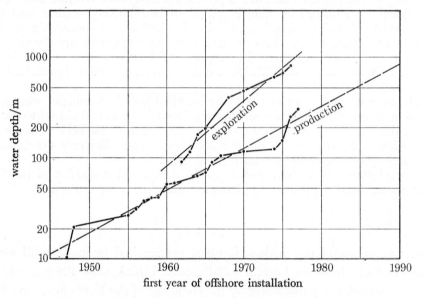

FIGURE 1. Maximum water depth for exploration and production.

Until comparatively recently it was generally believed that underwater well completions, floating production facilities and offshore loading and storage could only be justified in water too deep for fixed production platforms, the limit being 200–250 m. We have now, however, learned enough about the technology involved to appreciate that it can be applied not only to fields in deep water, but sometimes depending on the circumstances even more economically in shallower waters and with greater efficiency than conventional systems.

Figure 2, representing the development of a North Sea field, presents the problem diagrammatically. The production platform is shown in the centre of a circle and the black spots represent the wells to be drilled within it. You will see, however, that there are a number of wells which are well outside the reach of deviated drilling. These small and distant accumulations may, however, be susceptible to economic development by underwater completion techniques.

Two distinctly different problems are encountered here. The accumulation in the bottom left-hand corner calls for one producer at a distance of some 3.5 km from the platform. Production can be achieved by connecting the subsea completed well to the platform by means of an individual flowline.

The accumulation northeast of the platform is more complex and a total of four producers and three water-injection wells are presently foreseen. Moreover, the distance from the platform is approximately 6.5 km, which does not allow for individual flowlines because of excessive pressure losses in these lines. In this particular case use will have to be made of an underwater manifold which serves as collecting point for individual well streams.

Let us consider the two major building blocks that are required under the sea, e.g. the underwater well completion and the underwater manifold.

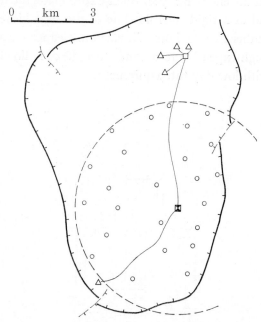

FIGURE 2. Use of satellite underwater completion wells and a subsea manifold in the North Sea. Quartered square, platform; ○, platform (deviated) producer; □, subsea manifold; △, underwater completion producer; ——, bottom laid flowline or bulkline.

One of the advantages of working in a major international community of companies is that a problem may be tackled in more than one way, the results compared, and a decision made which will serve the purpose best in a given set of circumstances. Research may also be shared with other groups working along different avenues towards the same goal. The size and spread of Shell companies' operations have enabled two separate approaches to be made to the problem of operating and maintaining subsea installations; by designing watertight air-filled compartments in which technicians can work in their shirt sleeves at the bottom of the sea; and alternatively by 'wet' subsea installations relying on remote control from the surface.

In 1972 research by Shell in the United States and Lockheed Petroleum Services led to the world's first ocean-floor well completion in a normal atmosphere wellhead chamber in 115 m of water in the Gulf of Mexico. Engineers sent down in a service capsule were able to assemble the wellhead equipment inside the steel chamber by using standard tools and techniques. Shell Oil has so far made nine underwater completions of which the earlier ones were of the 'wet' type.

On the other side of the world in Borneo, Brunei Shell Petroleum has completed ten underwater wells over the last nine years. A technique has been evolved by which subsurface work on the wells can be carried out by means of tools pumped down to the well through the crude oil flowline. Extensive tests on land have conclusively proved the feasibility of this method and many lessons have been learned on the reliability of submerged control systems (figure 3).

While the wells in the Gulf of Mexico were equipped for rather low flow rates of about 500–1000 barrels per day, the Brunei wells were designed for flow rates of 5000–10 000 barrels per day.

The technology acquired by separate but complementary research in two hemispheres is being coordinated and applied in deeper water and under more exacting physical conditions in the North Sea. Towards the end of last year Shell Expro made the first 'wet' well completion in the Brent field. Instead of a rigid steel flowline of the type hitherto used, a flexible high pressure steel flowline – rather like the hose of a vacuum cleaner – was devised to link the well with the platform. Although expensive to manufacture, flexible flowlines do not need pipelay barges and can be laid with low day-rate equipment.

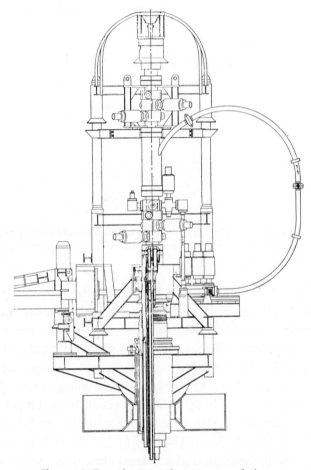

FIGURE 3. Brunei type underwater completion.

Let us now consider the second building block: the underwater manifold. Recently, field testing has begun in the Gulf of Mexico of a dry manifold centre which gathers, measures and controls production from three oil wells, two of which have been completed on the ocean floor.

Figure 4 gives an impression of an encapsulated well manifold centre. Its purpose is to serve as a collecting point – a sort of submarine flow station – to bring together the oil streams from a number of wells. The oil is then fed into a bulk line to the surface platform, thus avoiding a mass of costly flowlines and risers. Exxon has pioneered a wet manifold in the Gulf of Mexico at a depth of 54 m, and a French company has recently installed a prototype 'wet' manifold in shallow water offshore Gabon.

The next step is to design and build large manifold centres capable of handling the large daily flow rates expected from North Sea fields. The first of these (shown in figure 5) is being designed jointly by Shell Expro and the British Industry. The diagram gives no real impression of the size of the structure for which feasibility and engineering studies have been completed.

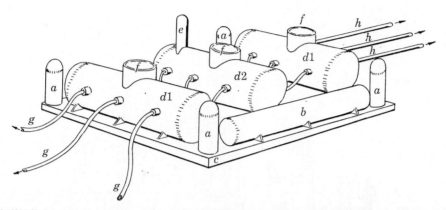

FIGURE 4. Shell Oil–Lockheed one-atmosphere dry manifold centre. The unit is floated to the site, sunk in position by flooding the ballast and trim chambers, and the flow, bulk, t.f.l. and control line bundles are pulled into the connection chambers, through sealing glands, by linear winches in chambers. (a) Trim tanks; (b) ballast tank; (c) raft foundation; (d) dry chambers; (d1) flowline, umbilical and bulkline connection chambers; (d2) manifold, instrumentation and test chamber; (e) test separator; (f) 'teacups' for mating of Lockheed transfer capsules; (g) flowline and control bundles to wells; (h) bulk, t.f.l. and control bundles to platform.

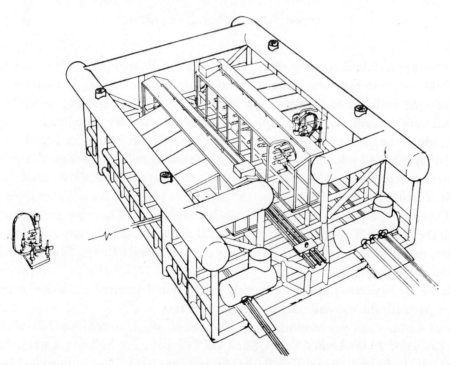

FIGURE 5. Shell Expro's proposed underwater manifold centre, accommodating both template drilled wells and tie-in points for satellite wells.

It will be about 48 m long, 42 m wide and 12 m high and weigh 1800 t. The main framework or 'template' will incorporate bays for eight well locations with facilities for connecting satellite wells by flowline to the manifold.

Some elements of the 'dry chambers' pioneered by Shell Oil and Lockheed in the Gulf of Mexico might be incorporated, facilitating easier connection, maintenance and repairs, thus reducing the need for divers or other sophisticated connecting methods. Shell Expro are planning several underwater well manifolds in the North Sea in their programme.

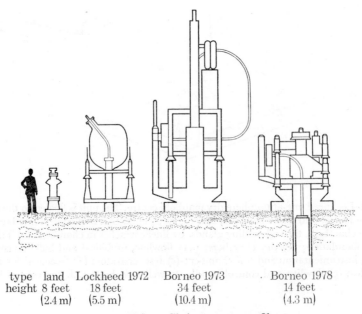

	type	land	Lockheed 1972	Borneo 1973	Borneo 1978
	height	8 feet	18 feet	34 feet	14 feet
		(2.4 m)	(5.5 m)	(10.4 m)	(4.3 m)

FIGURE 6. Subsea Christmas-tree profiles.

It is easy to appreciate that going under water calls for different technology and both wells and manifold are more expensive than conventional installations. One has only to glance at the sophisticated wellhead of the type used by Brunei Shell (see figure 6) to see how far we have moved from the comparatively simple Christmas tree of years gone by. It is appreciated that the height of the subsea wellhead could form a hazard if hit by trawler boards or dragging anchors. In an effort to minimize this risk, Shell International have designed a caisson type recessed tree concept which reduces the total height considerably and allows easier protection.

There is of course concern about the potential hazards of subsea installations from the point of view of operation, accessibility, inspection and maintenance. These aspects should be compared with the hazards of operation on conventional platforms. The factors affecting the safety of wells are, *inter alia*, size of clusters, probability of mechanical damage, fire risk and location of wells and process equipment.

On balance, we have concluded that the risks to which underwater installations are subjected to are on a par with the conventional facilities, if not less.

Of course, we have not yet reached the depth limit at which conventional fixed production platforms can work. In the United States, Shell Oil Company are building a steel platform for use in the Gulf of Mexico in more than 300 m of water (figure 7). The platform will be installed in three sections, one on top of the other, the first one of which will be placed this summer.

As can easily be appreciated, the depth capability of a floating structure is still much greater and the need for it might present itself soon. In April of this year Shell Teoranta began drilling an exploration well 220 km west of Ireland in 480 m of rough Atlantic water, using a Sedco-700 type semi-submersible. The well has now reached a depth of 3000 m.

I now turn to the various means by which oil can be produced from subsea wells and discuss the floating facilities required to process and handle it. What are the requirements to be met? First the well effluents must be taken up to the producing unit on the surface. Next each well might be tested individually to assess how much oil, gas and water is being produced. Then the treated crude, gas and water are conducted back via the bottom of the seafloor, either to an offloading point on the surface for oil or to an injecting point on the seabed (for gas or water).

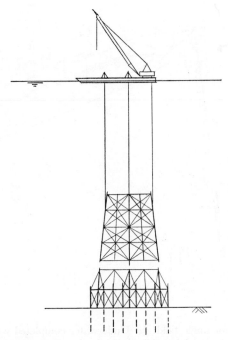

FIGURE 7. Multi-part platform (300 m water depth) for use in the Gulf of Mexico.

Hence a multi-functional number of conduits is needed, which is called the multibore production riser. Up and down and sideways movements of the surface facility with relation to the multibore riser have to be compensated for.

The first floating system installed in the North Sea was by Hamilton Brothers in the Argyll field, from which the first U.K. oil was shipped ashore in the summer of 1975.

A more recent development is the Single Anchor Leg Production System, S.A.L.S. (figure 8), for underwater wells. In this arrangement, a single anchor leg composed of chain links is connected from a universal joint at the bottom to a top swivel on the surface incorporating separate concentric flow paths for oil and for control hydraulics. A tanker is used as a floating base for production facilities and storage and the unit is able to swing on its axis in response to wind and wave forces. It is equipped with oil and gas separation equipment and facilities. The gas will be used as fuel in the ship's furnaces or burnt in specially built incinerators. Oil is loaded into a shuttle tanker which fills up alongside and takes it ashore. The first S.A.L.S. unit to be put to operational use was due to start work in June 1977 in an oilfield in 115 m of water east of Barcelona in Spain.

To overcome some of the limitations inherent in the single anchor leg system, we are studying a more complex layout which incorporates a bundle of conduits between the floater and the fixed manifold (illustrated in figure 9). Here there is a template manifold centre (*a*) on the seabed capable of accommodating six wells, with facilities for connecting satellite wells (*b*) through flowlines. The oil comes up through the production riser (*c*), which incorporates a number of flowlines and one 'export line'. The semi-submersible platform on the surface (*d*) has the necessary production facilities. After processing, the oil is pumped back through the manifold for loading into the tanker via the export line (*e*) and the single anchor leg (*f*) into the storage vessel, to which a shuttle tanker (*g*) can be moored. This system is suitable for use in deeper water.

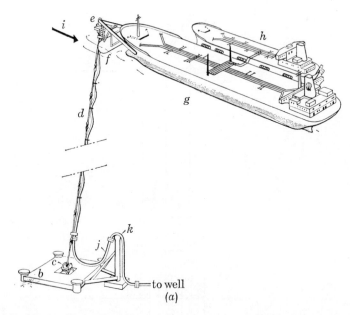

FIGURE 8. 'S.A.L.S.' installation for single field. (*a*) Underwater completed well with wet Christmas tree; (*b*) four-pile base; (*c*) bottom universal joint; (*d*) Single Anchor Leg composed of 'bicycle' type chain links, each approx. 16 ft long and 8 in in diameter; (*e*) top swivel with separate concentric flow paths for oil and for control hydraulics; (*f*) mooring 'A' frame with cylindrical flotation tank to maintain anchor leg in tension; (*g*) floating unit storage (f.u.s.) tanker with oil/gas separation facilities; gas is burned in ship's furnaces and in a closed incinerator; flaring is not permitted; (*h*) shuttle tanker carries oil to shore; shuttle may now moor alongside in calm environments but will moor in line from bows in rougher weather areas; (*i*) f.u.s. tanker is free to 'weathervane' but can be moored by her own bow anchors over the well to allow wireline maintenance of the well; (*j*) 'Coflexip' flexible flowline carries the well stream to the surface via swivel (*e*) and jumpers over 'A' frame; (*k*) conventional steel flowline.

It will be appreciated that we are moving by gradual evolutionary steps to improve our grasp of subsea engineering technology. The manifold centre development is a great advance towards eliminating a clutter of steel spaghetti on the seabed, apart from the operational advantages it brings. One of the more interesting techniques is the one known as 't.f.l.' (through the flowline) maintenance of wells. In the system I have just described it will be feasible to reach the wells from the platform without the help of divers by passing tools through the flowline using a hydraulic circuit. The tools are mounted on what I might call a 'hydraulic locomotive' and by selecting the proper tool subsurface valves in the well-bore can be operated or exchanged. Operations can be monitored from the surface. Unlike the conventional method of running tools down on a wire line vertically, for which a semi-submersible is required, the hydraulic system will be able both to 'push' and 'pull' the tools and to enter the well from a remote point.

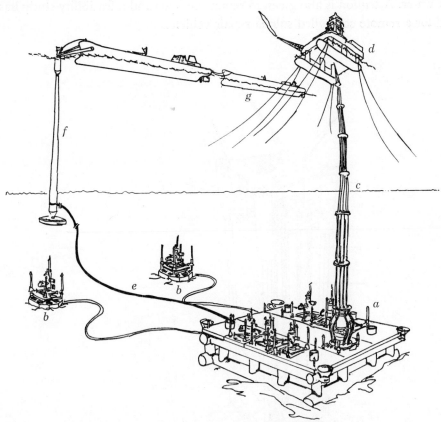

FIGURE 9. Floating production facility with subsea completions. (*a*) Six-well completion template; (*b*) satellite wells tied to template by flowline; (*c*) production riser: eight flowlines and one export line; (*d*) semi-submersible with production facilities; (*e*) export line; (*f*) S.A.L.M. with A-frame-moored f.u.s. tanker; (*g*) shuttle tanker, stern moored to f.u.s. tanker. Well maintenance is by wireline.

Figure 10 shows a still more advanced means of producing oil from subsea wells by a floating platform anchored vertically by tension lines which hold it in place. The platform is moored securely to the bottom base-plate with the multi-riser system enclosed within it. Vertical motions are supressed due to the constant distance between the platforms and the anchoring point.

I should now like to mention surface storage systems. The Shell-designed storage and loading facility known as Spar consists of a floating cylinder rather like an enclosed inverted wine glass some 100 m high, and with a diameter of 30 m, able to hold 3×10^5 barrels of oil. Pending the completion of a pipeline system to shore, the Brent field is being initially produced through a Spar installation. Tankers are loaded through flexible hoses, with the oil fed into the Spar from a pipeline manifold on the seabed connected to the producing platforms. We are now studying the possibility of developing a semi-Spar production, storage and offloading system, linked with a manifold centre on the seabed. The concept is illustrated in figure 11.

The systems which I have described are to a certain extent dependent on sea-bottom pipelines. We are now laying large diameter pipelines in 150 m of water and recently a 16-in pipeline was laid in 550 m of water in the Straits of Sicily as a test. Furthermore a joint-industry effort (with 37 participants) is well on its way to develop methods for laying 30-in pipelines in 1000 m of water. Attention is also given to repair methods and a feasibility study has just been concluded for a remote controlled subsea repair vehicle.

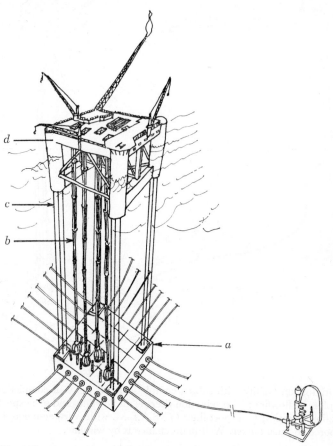

FIGURE 10. T.P.P. production system. (*a*) Bottom baseplate; (*b*) riser system; (*c*) mooring system; (*d*) T.P.P. unit.

As pipeline laying methods gradually develop into underwater construction, hyperbaric welding techniques will be improved upon. Experiments with this technique have proved the feasibility in 300 m of water and this method is now routinely used to make pipeline tie-ins in the North Sea in 150 m of water. Hyperbaric welding techniques are of course necessary if bottom-tow construction methods are to be developed to full maturity.

In carrying out all these exacting tasks in extreme physical conditions, one can hardly overestimate the importance of precise weather monitoring equipment, which is vital in assessing design criteria and the safety and reliability of the equipment. In the exploration phase the annual fair weather season could be selected for drilling operations; in the production phase the weather must be coped with all the year round. Hence weather buoys have been developed which collect data (including profiles of currents from the surface to the sea bottom), and transmit information at regular intervals.

The combined efforts of Shell International and Marine Exploration Limited (Marex) have resulted in the design of an oceanographic data gathering buoy (figure 12) for remote and exposed areas. A prototype was tested in 1975 in the Celtic Sea and a self-recording version was launched in the Atlantic north of Scotland last December. It has functioned well even in the extremely bad weather of last winter, and is providing data about an area of which limited detailed environmental information is available.

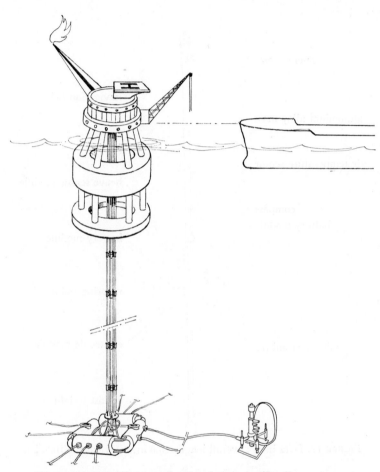

FIGURE 11. Semi-Spar production (storage) offloading system.

I have briefly mentioned the individual components under study. What we are now working towards is the conceptual design of complete production systems, incorporating subsea completed wells, manifolds, risers and floating units. There are relatively small oil accumulations where the provision of additional production platforms at enormous cost could not be justified, but where a combination of subsea wells and manifold with a floating facility on the surface might, be economic.

The offshore storage and loading terminal has also proved its practicality, although problems do still occur.

I hope that it will not be felt that I have been leading a guided tour through a Technological Wonderland under the Sea or indulging in a Jules Verne fantasy. There is nothing visionary or speculative about the technology which now exists. It *does* work. There are now nearly 30 producing Shell wells on the seabed in various parts of the world.

Taking into account the steep rise in the price of all forms of energy, and hence oil's ability to bear higher technical costs than in the past, there are grounds for confidence that subsea engineering can provide technological solutions to development problems which can be economically applied. I am convinced that floating systems will play an important rôle in deep water in the future and provide a competitive alternative to conventional platform development in shallow water.

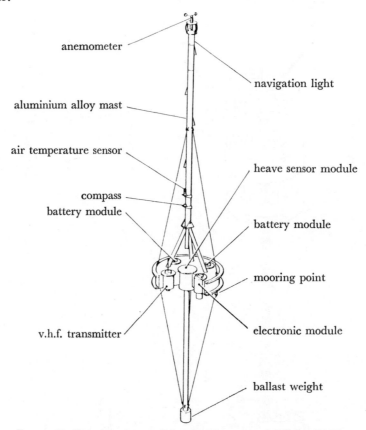

anemometer

navigation light

aluminium alloy mast

air temperature sensor

heave sensor module

compass
battery module

battery module

mooring point

v.h.f. transmitter

electronic module

ballast weight

FIGURE 12. Data buoy: overall length 16.6 m; total weight 1000 kg.

Discussion

T. S. McRoberts (*Q.M.C. Anchor Technology Ltd., 229 Mile End Road, London E1 4AA*). Mr Goldman has given us a broad sweep of subsea technology and its problems. It raises a number of questions but I shall confine myself to one in which I have a personal interest. The anchoring problems are likely to be very considerable as we move into deep waters and more and more of these structures have to be floated, perhaps in groups. Could he tell us a little about any special problems he can foresee in deep waters?

E. C. Goldman. Anchoring problems have already been encountered in shallower water, e.g. dragging of conventional type anchors, use of 'piggy-back' anchors or drilled-in anchor piles to cope with adverse soil conditions, wear and fatigue of chain and wire rope. Moving into deeper water will aggravate these problems.

In particular the seabed mooring points are a point of concern. The tendency of using wire rope with reduced lengths to minimize the weight to be carried by the floater will cause a vertical component of the anchoring force to act on the anchors particularly during major storms. Such considerations necessitate additional corrective measures.

Conventional types of anchors are not suitable and therefore supplementary means should be developed. In this connection assistance from propulsion units to safely ride out major storms could be a good engineering solution. Such a solution gives the possibility of reducing the offset of the floater which is likely to be required because of the weak restoring force characteristic inherent in deep water anchoring systems.

J. BLACK (63 *The Woodlands, Esher, Surrey, U.K.*). What is the maximum sea-state which allows a tanker to moor at a Spar buoy? Please describe the method used by shipboard personnel to pick up the floating mooring hawser in such a sea condition, bearing in mind the importance of tanker turn-round time and the need to prevent damage to the Spar mooring.

E. C. GOLDMAN. The present, although still early, experience in mooring-up of tankers to Spar is that the procedure is still safely performed in seas of approximately 12 ft significant wave height and wind speeds of up to 25 knots.

The applied and well-proven method is basically that the shipboard personnel uses a grapnel, or in some cases an airgun, to recover a nylon messenger line of 3 in diameter and 1000 ft long, which is streamed out from the Spar upon arrival of the tanker. The messenger line is put subsequently on the ship's winch, and kept reasonably taut by heaving it in while the tanker is slowly approaching Spar. Both the main mooring hawser and loading hose connection wire are attached to the end of the messenger line. When the chafing chain of the hawser has come on board and been secured on the stopper, the hoses are pulled across and connected to the ship's manifold and loading can commence.

During the mooring the ship gives 5–10 t astern propulsion depending on the sea condition. Quick-release devices are incorporated to protect the mooring and hose connections. The total procedure in extreme conditions has taken a maximum of $1\frac{1}{2}$ h.

Phil. Trans. R. Soc. Lond. A. **290**, 113–124 (1978) [113]
Printed in Great Britain

The challenge of producing oil and gas in deep water

By W. H. van Eek

Delft University of Technology, Division of Mining Engineering,
Mijnbouwstraat 20, Delft, The Netherlands

The paper outlines the present state of the art of deep sea drilling and discusses some of the problems with the 'controlled' as well as the 'uncontrolled' techniques.

The first method is being developed by the contract drilling companies under the auspices of the oil industry, while the second method was introduced by the Deep Sea Drilling Project under the guidance of a group of oceanic institutions (Joides).

The oil industry has drilled controlled in water depths of up to 1000 m and slightly over, and is now capable of extending the technique to 2000 m. Joides contemplates controlled drilling in water depths to 3650 m, say by the end of 1981.

It is suggested that not only the slope but also the rise of the continental margin should soon be investigated in a number of suitable localities in order to assess adequately the potential of the last remaining major unexplored frontier for oil and gas.

The paper emphasizes that it is already possible today to carry out controlled exploration even to water depths of over 4000 m. If such exploration were successful, production could also be achieved by making use of the presently developing underwater technology in 200–300 m of water.

1. Introduction

Although it is at present difficult to obtain concessions or drilling permits from individual countries in deep water, especially in view of the uncertainty with regard to ownership, it is hoped that the third 'Law of the Sea' conferences may soon reach an agreement so that the position is clarified. This paper assumes that within a few years exploration on the continental margins will not be unduly restricted even in the absence of international legislation.

In order to assess the possibility of finding hydrocarbon accumulations of sufficient interest in deep water, modern seismic techniques are of great value, especially as not only structures or other traps are located, but even the presence of hydrocarbons may be revealed. However, this feat is at present only possible if a few stratigraphical tests or key wells can be drilled in such areas. Moreover, today it does not seem very clear where on the margin the best prospects for finding large accumulations of hydrocarbons may exist, a statement which calls for a short explanation in appendix 1 of this paper.

It may therefore be of prime importance that such key wells be placed over the total width of the continental margin in order to assess the area in one sweep. This means that also some key wells are to be drilled in water depths of up to 4000 m or perhaps even somewhat deeper. However, such efforts are of little economic interest, if no assurance can be offered, once valuable oil or gas accumulations are discovered, that economic production is also feasible. The trends in past and present developments of off shore drilling and today's underwater production techniques are therefore of particular interest in order to forecast whether in the near future adequate assessment of the deeper offshore could be undertaken, and, if successful, followed by production.

2. PAST AND PRESENT DEVELOPMENT IN CONTROLLED OFFSHORE DRILLING

The petroleum industry completed by the end of 1965 two decades of offshore drillship experience. In 1965 an exploratory well was drilled in 192 m of water by the Exxon Corporation in the Santa Barbara channel off the coast of California (Anon 1977). This record was achieved by means of so-called 'controlled drilling' commonly practised by the industry in the offshore. This method is based on techniques developed for land drilling and adapted to a restless and forever changing sea environment. Technology to counteract the motions of the sea has been improving steadily and by 1965 exploration up to the edge of the continental shelf was feasible.

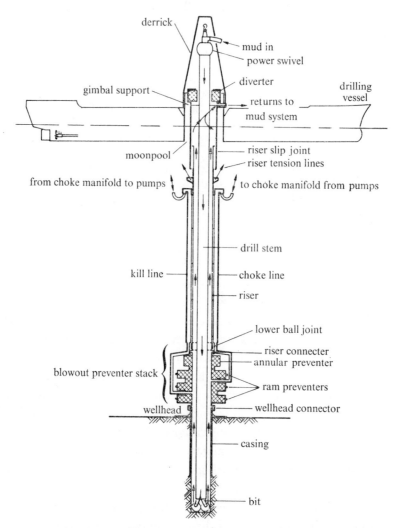

FIGURE 1. Riser drilling system. Guidelines, buoyancy, subsurface buoys, etc., are not shown.

The control in controlled drilling is exercised by a circulating mudflush of a specific gravity sufficiently high to ensure that the pressure inside the borehole opposite a permeable layer is always in excess of the fluid or gas pressure in the layer itself. The use of the mud is only economical if the mud is returned to the vessel; for this purpose a conductor pipe or riser is used which is firmly fixed to the wellhead connections at the sea bottom and flexibly attached to the drill vessel. The riser should have a diameter large enough to allow the passage of the largest drill bits. The procedure is as follows: the mud is circulated downwards through the drillpipe and returns upwards to the drillship through the annular space between the riser and the drillpipe. The horizontal movements of the drillship under normal conditions are sufficiently restricted by its anchoring that the riser cannot be snapped off at the seabed. Moreover, a ball joint is installed at the bottom of the riser and helps to prevent excessive bending (figure 1; McLeod 1976). Finally the drillship's vertical movement or heave can now be compensated satisfactorily for the drillpipe as well as for the riser. The riser has also to be tensioned in order to prevent buckling which is not only caused by its own weight, but also by the difference in specific gravity of the mud inside and the seawater outside the riser. It is easily understood that the deeper the water, the more difficult it becomes to satisfy these requirements.

In case of a threatening blowout when the mud does not adequately equalize the pressure of one or more layers below the seabed, the well can be controlled by closing a blowout preventor valve at the seabed. This is possible if a so-called blowout preventor stack is installed consisting of a number of valves equipped with hydraulic rams which can, for instance, close around the drillpipe. Some type of electro-hydraulic system should be installed together with the riser in order to operate the valves from the drillship.

Once the well is closed in, remedial action should be taken and for this purpose two additional small diameter high-pressure lines need to be placed alongside the riser, the so called choke and kill lines. Such arrangements again add to the technical complications of the riser design, especially as the water becomes deeper.

The industry continued improving its riser design. Shell drilled in 1975 an exploratory well in 700 m of water offshore Gabon, while a year later Exxon drilled a deep test in 1060 m of water offshore Thailand.

The inclusion of buoyancy into the riser design is one of the major improvements which made it possible to achieve these records. The negative buoyancy of the riser is nearly compensated for by placing around the riser in concentric rings so-called syntactic foam made by Emerson & Cumming Inc. (Watkins & Howard 1976), consisting of hollow glass spheres in a binder of epoxy or polyester resin, having a density of about 500 kg/m³. The material is now proven to withstand for a considerable time pressures equivalent to 2000 m of water, and it is expected that in the near future it can be used to 3000 m. Another system to achieve, say, 95 % buoyancy of the riser is provided by Regan and consists of a number of floats which are regularly spaced along the riser and which can be emptied by air pressure.

Today the weight of a 16–18⅝-in riser can be adequately compensated to a water depth of 2000 m. The riser should only be tensioned to take care of the differential weight between mud and seawater and the dynamic forces exerted on the riser by waves, current and surge of the vessel. Nevertheless, the pull to be applied to a 18⅝-in riser may well exceed 5000 kN when using heavy muds in bad weather in 2000 m of water. The latest 16–18⅝-in riser designs are capable of use in 2000 m of water.

Other necessary techniques of drilling in deep water which are applied by the oil industry are discussed in the next section, as these were initiated or developed by the Deep Sea Drilling Project (D.S.D.P.) (Peterson 1975).

3. Past and present development in uncontrolled offshore drilling

In the summer of 1965 a proposal was submitted by Scripps Institution of Oceanography, University of California at San Diego, to the National Science Foundation for 'Drilling of Sediments and Shallow Basement Rocks in the Pacific and Atlantic Oceans and Adjacent Seas'. This proposal lead to the formation of Joint Oceanographic Institutions Deep Earth Sampling (Joides). It was decided to initiate the D.S.D.P. under management of the Scripps Institute, the project being financed by the Federal Government of the U.S.A.

In the summer of 1968 the drillship the *Glomar Challenger* began its first round of drilling in the deep sea. The ship was equipped with a dynamic positioning system to keep the ship on station, without anchors. This is achieved by means of additional propellors suitably placed alongside the ship. The position of the ship *vis-à-vis* the seabed is determined acoustically, and signals are translated by means of a computer into the appropriate propeller response.

Many holes were drilled in water depths of over 4000 m and even up to 7000 m during the project, but no riser was used. The bit on the drillpipe enters the seabed, and seawater is circulated through the drillpipe. Drill cuttings do not return to the vessel, but spread out over the ocean bottom. Continuous coring is therefore required and is carried out by means of wireline coring, the core barrel being extracted and reintroduced through the drillstring. This method is called uncontrolled drilling, as no remedial action can be taken against threatening blowouts apart from abandoning the hole by means of placing a cement plug.

In the beginning, trouble was experienced if the bit encountered hard, abrasive streaks of chert. Once the bit was dulled, the hole had to be abandoned. A breakthrough was established by developing a re-entry method. A large funnel-shaped cone is pushed into the seabed, and by means of sonar signals can be relocated with the drillstring. The drillstring extends almost to the rim of the cone, and the ship is slowly moved across it; once the drillstring is directly above the cone, the bit is at once lowered into the cone. The technique is now so well perfected that little time is lost in relocating the hole with a new bit. This method or similar ones are also used by the oil industry before the placing and cementing in of the first casing string. Once this operation is carried out, the blowout preventor stack will be installed together with the riser, and uncontrolled drilling becomes controlled. In 1975 the D.S.D.P. was converted to the International Phase of Ocean Drilling (I.P.O.D.). France, Germany, Great Britain, Japan and Russia participate with the U.S.A. in this project. Recently discussions were started as to whether the project should be continued for a considerable number of years and if so, whether the vessel selected for scientific ocean drilling should be equipped with a riser system.

It is possible that in view of availability of suitable equipment for ocean depths up to 7000 m, the riser problem could be solved in a manner rather different from the one preferred by the oil industry. If the I.P.O.D. project is continued, the experience of controlled drilling in water depths of up to 7000 m may be of great interest to the petroleum industry.

4. Pressure control of wells drilled in great water depths

One could wonder why such large risers of a diameter of 16–18⅝ in need to be used at all. The pull required to prevent a 10¾-in riser from buckling is about half that required for a 18⅝-in riser under similar conditions, because the 10¾-in riser contains much less mud, while the dynamic forces are also considerably lower. The apparent extravagance of using large risers is, however, caused by the need to balance the pressure in the borehole in such a way that on the one hand the pressure opposite the formation is high enough to prevent influx of fluids or gas into the hole, but on the other hand low enough to avoid fracture initiation (Koch 1975). In figure 2, pressures are plotted against depth for the case where the seabed is at 6000 ft (approx. 1830 m).

The author regrets that the petroleum industry on the whole is not yet using the International System of Units (SI). He personally made a strong plea during the Petroleum Industry

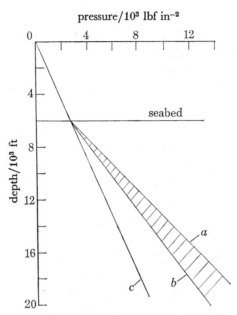

FIGURE 2. Pressure–depth relations: seabed at 6000 ft: line a, 1.0 lbf in^{-2}/ft; line b, 0.815 lbf in^{-2}/ft; line c, hydrostatic gradient of 0.445 lbf in^{-2}/ft.

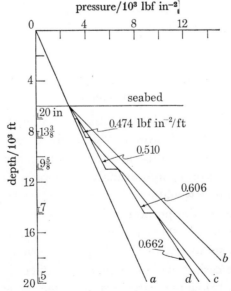

FIGURE 3. Pressure–depth relation for selected casing scheme: seabed at 6000 ft. Line a, hydrostatic gradient line (pore pressure trend); line b, overburden gradient line; line c, formation fracture gradient line; line d, maximum mud gradients acceptable.

Seminar held by the United Nations Environment Programme in Paris, March 1977, to select at least one standard system of units applicable to the well-control programme to avoid calculation errors (van Eek 1977). For the present it seems less appropriate to use SI units in view of the industry's practices.

Figure 2 shows the highest liquid (gas) pressure gradient (a) of 1.0 lbf in^{-2}/ft which one can expect below the seabed, as compared with the hydrostatic gradient (c) of seawater of 0.445 lbf in^{-2}/ft. If a pressure gradient of 0.815 lbf in^{-2}/ft is reached below the seabed (b) the possibility exists that the formation may fracture due to excessive overpressure of the mud in the borehole. This can for instance be initiated at 12 000 ft when the mud gradient from sea-level amounts to only $\frac{1}{2}(0.445 + 0.815) = 0.63$ lbf in^{-2}/ft. Controlling the formation pressures from sealevel is therefore the more difficult, the deeper the water depth becomes.

Figure 3 shows the following relations:

(a) The hydrostatic gradient line shows the relation between depth and pressure, if the pore pressure trends hydrostatically.

(b) The overburden gradient line shows the maximum pore pressure related to depth which may be expected. This pore pressure then equals the overburden pressure, and any higher pressure would enable the pore contents to lift the overburden.

(c) The formation fracture gradient line shows the fracture pressure related to depth. Any higher mud pressures in the boreholes are likely to fracture the formation vertically. This relation being only valid for hydrostatically trending pore pressures, is derived as an average of many field studies.

(d) The discontinuous line indicates the maximum mud gradients acceptable in various intervals between successive casing shoes. One runs the risk by exceeding this gradient that the formation will fracture near the casing shoe of the last set casing. The remaining triangles indicate for which pore pressures the maximum allowable mud gradient is insufficient to control the well. Obviously, the interval chosen between casing shoes depends on the risk one wishes to assume, but it is clear that one cannot eliminate this type of risk altogether, unless one would set casings continuously. It is also evident, if one has to set casings frequently for protection, that the advantage of starting with a large marine riser is rather quickly lost.

5. ALTERNATIVE METHODS TO DRILL WELLS IN WATER DEPTHS TO 4000–5000 m

Large risers have recently been designed for a water depth of up to 2000 m and may eventually be used in these water depths. However, the author does not visualize that these industrial designs can soon be scaled up for use in water depths of around 4000 m. It is therefore interesting to see whether, based on today's available technology, alternative methods can be proposed which in the near future would enable prospecting up to such water depths.

(a) It is now suggested using for this purpose a riser of diameter $10\frac{3}{4}$ in because the tensional forces to be applied to this riser in 4000 m water would be no more than those exerted on a 2000 m $18\frac{5}{8}$-in riser. Figure 4 shows how one could possibly overcome the problem of having to set a number of protective casings without having to reduce the hole diameter so quickly that a reasonable penetration below the seabed can no longer be expected.

For instance, a $10\frac{3}{4}$-in conductor is set 1000 ft below the seabed, and the underwater blowout preventor stack together with a $10\frac{3}{4}$-in riser are attached to this conductor. Subsequently a

$9\frac{5}{8}$-in hole is drilled to some 3000 ft below the shoe of the conductor and a 7-in casing is set at the bottom of the hole. Now a $5\frac{7}{8}$-in hole is continued some 3000 ft below the 7-in casing shoe. If neither unexpected pressure increases nor hydrocarbons are found, the 7-in casing is cut just below the shoe of the $10\frac{3}{4}$-in conductor. After recovery of the free part of the 7-in string, the hole is sidetracked and a new $9\frac{5}{8}$-in hole is drilled to the depth which was reached by the $5\frac{7}{8}$-in bit and a 7-in casing is again set. The protection of the 7-in casing is extended another 3000 ft. This procedure could be repeated several times, which may be necessary if one is forced to select short intervals between successive casings.

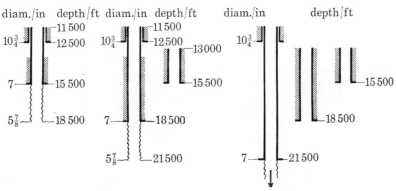

FIGURE 4. Sidetracking method avoiding loss of hole diameter when setting protective strings.

The drawback of this procedure is obvious; the hole may have to be redrilled several times and sidetracking is costly. The method becomes attractive if the technique is only required in certain holes which are planned for deep penetration. It allows the use of a smaller ship, a more manageable riser pipe, and above all, it may provide right now a technique of drilling holes in water depths of over 13 000 ft (4000 m).

(b) Another alternative is the use of an underwater float, or buoy, sometimes called a pedestal.

The underwater buoy aims to place the underwater blowout preventor stack just deep enough so that the dynamic forces are greatly reduced, although the current influence can never be eliminated completely. In this manner a long riser is avoided, and the wellhead is within the diver's reach. Moreover, in extremely bad weather when sufficiently exact positioning of the drillship is no longer possible, the riser needs to be disconnected from the underwater stack at the seabed. Drillpipe and riser are then usually pulled. The longer the riser, the costlier such an operation becomes. In the past this system was rejected.

First, it just did not seem possible to design a conductor between the seabed and the underwater float which was able to withstand the pull needed to compensate the static forces as well as the dynamic forces caused by the drift of the buoy due to the assumed underwater currents; secondly, failure of the conductor between seabed and float (its weakest point being at the float) would endanger the drillship, as the positively buoyant float could hit the vessel with tremendous impact.

(1) D. Meyer-Detring (1976) describes the use of a disconnectable floating riser carrier (figure 5). This project was undertaken by a number of German industrial firms. The patent claims that the use of this float would overcome the extraordinarily high cost of uncoupling and pulling a 2000 m riser, equipped with buoyancy material in heavy weather. It works as

follows: during bad weather the floating riser carrier is disconnected from the drillship and pulled by winches installed on the float to about 200 m under the water surface. The shortened riser (by means of a telescopic construction) is kept under tension by the tensioners of the float. The float itself is dynamically positioned. Presumably the float's positive buoyancy equals the tension which has to be pulled on the riser. It is then kept by the riser itself in place. Because the ship is only connected by an umbilical cable to supply energy to the float, it can move away from the float, so that in case of riser failure the ship cannot be damaged. The riser carrier is again coupled back to the mother unit when weather conditions permit to resume operations. By equipping the float with dynamic positioning, and safeguarding the ship against riser failure, the previous objections are overcome.

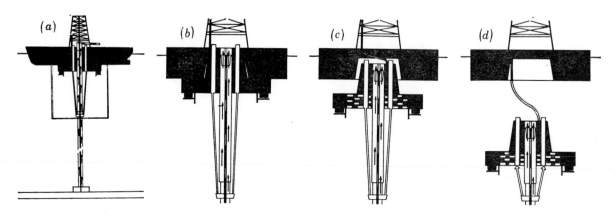

FIGURE 5. Disconnectable floating riser carrier. (a) General view. (b) Phase 1: disconnection. Enlarged section of (a). The drill string is fastened to the riser carrier, the flow line is replaced by a circulating pump. (c) Phase 2: flooding. The riser telescoping joint is pushed together about 3 m. The umbilical cable is connected with the riser carrier. (d) Phase 3: diving. The tensioners have pulled the unit about 200 m below the water surface. The telescoping capacity of the riser is reached.

(2) As an alternative to having a float which can be lowered from the ship, a permanent pedestal could be installed which also avoids the previously mentioned objections.

Although many designs of the pedestal are possible, the pedestal chosen in this paper (figure 6) consists of a number of large diameter pipes and flotation chambers at its lower end. Total height is about 100 m. The pedestal is towed in a horizontal position similar to the way large jackets are now being floated out. By properly manipulating the flotation chambers the pedestal is placed at the drilling site in an upward position. Thereafter it is made slightly negatively buoyant, but supported at some 15–20 m below the sea surface by means of some floats at sea level. The drillship is now placed over the pedestal and drills through the pedestal. The pedestal is kept in place by dynamic positioning. The drillstring is drilled a 300 m into the seabed. Before cementing, the drillpipe pushes the pedestal to 200 or 300 m below sea level and is then cemented in. After the cement has hardened, the pedestal is given some positive buoyancy, being kept in place by the drillpipe. The drillship can now disconnect the drillpipe at the pedestal. The main hole is then drilled through the pedestal and the conductor cemented in. Now by making the pedestal sufficiently positively buoyant, it will exert the necessary pull on the conductor. In case of conductor failure the ship is protected, because the drillpipe will still hold the pedestal in place. It speaks for itself that drillpipe and conductor below the pedestal will also make use of syntactic foam or apply the Regan's flotation cans. In this manner the pedestal does not need to carry more than the excess mud weight, while the horizontal forces are taken care of by the dynamic positioning system. Figure 7 compares the conventional method with this alternative.

(3) In appendix 2 a slim hole drilling method is described, which could eventually be an attractive alternative, provided field tests prove the soundness of the research.

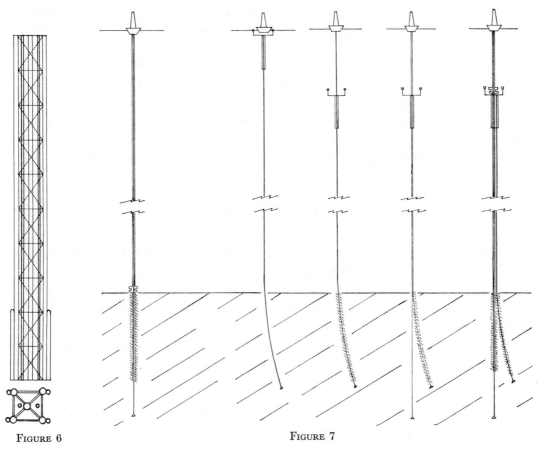

FIGURE 6   FIGURE 7

FIGURE 6. Sketch of suggested pedestal design.
FIGURE 7. Conventional controlled drilling method versus pedestal operation.

6. PRODUCTION TECHNOLOGY IN DEEP WATER

Production platforms like the ones set in the North Sea may be set in water depths of up to 300 m or even a little deeper within the next few years. However, the industry is convinced that in real deep water underwater completions have to be used. So far, underwater completions are installed in isolated instances, and although these completions have been successful in water as deep as 100–150 m, their use is still considered experimental by industry. The paper by Goldman (1978, this volume) covers the technological progress which has been made in the last few years. Perhaps one is now allowed to assume that soon underwater completions can successfully compete with large production platforms in water depths of around 150–200 m.

During the next five to ten years a number of offshore fields may be developed mainly by underwater completions, and further experience will be gained in offshore fields in up to

200 m of water. Having said this, it is thus the present lack of experience which slows down the incentive to explore in water much deeper than 200 m.

However, by making use of dynamically positioned production pedestals placed 200 m below sealevel and being equipped for a cluster of six to eight producers, today's technology could be used. For instance, the wells could be deviated below the seabed, like the producers now being drilled directionally from production platforms. A dynamically stationed spar or tanker-hulk could be installed above the pedestal, not only for the purpose of oil storage, but also equipped with the necessary production facilities. It would provide a source of energy for the pedestal.

As a matter of fact, even if wells could be produced from 2000–4000 m by underwater completions at the seabed, it is questionable whether a pedestal would not provide a better answer. In any case it would avoid the long and tricky connecting of flowlines between production facilities at the seabed and the floating storage at the sea level. It will also allow diver's access. A study was made to compare both production methods, while assuming that either one would be feasible technically. It was found that in either case the same optimum production could be achieved.

7. Conclusions

(a) It is believed that within the reach of today's technology the industry can drill and produce hydrocarbons in water depths of up to around 4000 m, provided the production per well is large enough to justify such a project economically.

(b) In view of the looming long-range shortage of hydrocarbons the petroleum industry should be encouraged to commence testing of the deeper offshore over the entire width of the margins in a number of suitable localities.

(c) The hope is expressed that governments will encourage the petroleum industry by making prospecting of the margins attractive.

Appendix 1. A justification to look not only on the continental slope but also on the continental rise for hydrocarbon accumulations

The *Glomar Challenger* found on leg 48 (*Geotimes* 1976) large hiatuses in the stratigraphical series above the Aptian–Albian, the upper Lower Cretaceous, at the edge of the margin in the Bay of Biscay. It is noted that the hiatus in the Upper Cretaceous is contemporaneous with the well-known global transgression, although its origin is uncertain. It is thought by the authors of the article in *Geotimes* that the relation is unrelated to the subsidence of the margin. They suggest, however, that the effect of the transgression was to concentrate carbonate production in warm shelf seas, depleting the oceanic water of carbonate.

One may perhaps speculate in view of the global nature of the Upper Cretaceous transgression that the environment along the width of the passive margin may have been sufficiently favourable to allow the formation of abundant carbonate source rocks in a number of localities before the major subsidence of the margin took place. If this had been the case, then generation of hydrocarbons could have been initiated after sufficient subsidence and coverage by sediments; and although vertical migration should not be excluded, horizontal migration could have played a more important rôle.

In the article 'The habitat of some oil' Knebel & Rodriguez (1956) reported their study of a large number of basins and showed that in general, hydrocarbons migrate preferably towards the stable part of the basin.

The gentle slope of the formations underneath the continental rise away from the edge of the margin seems to resemble the favourable conditions for migration towards the stable part of the basin as described in the article. It is therefore suggested that preferential migration towards the edge of the margin could have occurred in a similar manner, the hydrocarbons being trapped *en route* under favourable circumstances.

It could even be imagined that hydrocarbons might be found in stratigraphical traps in the sediments overlying the ocean crust, provided, for instance, at least some sediments with reservoir characteristics were deposited during important regressions. Although the subject can only be touched briefly in this paper, there seems to be every reason to consider these areas as prospective.

APPENDIX 2. A SLIM HOLE DRILLING METHOD

It is suggested to use a slim hole drilling method by means of diamond bits which can make progress through unconsolidated formations because the temperature of the circulating mud is minus 21 °C. The method of drilling with fluids which freeze the formation on contact is not new and has been used in several projects in order to improve coring recoveries in unconsolidated formations. However, the method has only been applied in shallow boreholes. The circulating fluid normally consists of diesel oil. The cooling power carried by the circulating fluid in order to freeze the fluid contents of the formation should be delivered at a sufficiently low temperature. Research was undertaken in Delft on the possibility of carrying eutectic ice in the mudstream in order to keep the temperature at minus 21 °C. The advantages are manifold. In the first place, clays will be frozen while drilling with the diamond bit so that progress will not be hampered. In the second place, it is hoped that the borehole will be sufficiently round and not oversized, so that just above the drill collars one or more packers can effectively be set in case of penetrating a formation with pressures higher than usual. In other words, blowout prevention takes place as near as possible to the fluid entry. The density of the fluid should be so low as to be slightly lower than the hydrostatic head which is expected to be present in permeable formations. This can be achieved by mixing kerosine or diesel oil with a solution of sodium chloride of a eutectic composition. The formations will not be entered by the fluid with high salt content, thus preventing freezing. On the contrary, just above the bit, freezing will commence and the borehole walls will be sealed off.

Continuous coring will be done through the centre of the bit, the cores will be ejected at the side of the bit and carried to the surface in the annular space by the mudstream. At the surface a certain amount of backpressure will be applied in order to keep the borehole under pressure. The bit is preferably rotated by a mud motor, like for instance a dynadrill. In case hydrocarbon-bearing formations with high pressures are encountered, their presence should immediately be revealed by a considerable outflow of the mud and the packer should then be set. Pre-prepared heavy mud can then be circulated above the packer in order to control the well or alternatively squeezed into the formation. In emergencies, cement can be placed above the packer. As the circulating fluid used should normally be of a lower density than the seawater, the riser pipe which is attached to the conductor should not need any more buoyancy then it already posesses due to the presence of syntactic foam (which, by the way is also an excellent thermal insulator).

Before disconnecting during bad weather, seawater will be circulated and the drillpipe pulled. The method is practically worked out as far as the initial research is concerned and a drilling test should now be programmed in the first place for a shallow hole onshore, say a few thousand feet deep. The calculations showed that the method can be employed at least up to 10 000 ft of formation which should be ample for reconnaissance in the deep offshore. A patent has been applied for and it is hoped that in the near future Delft can find sponsors to try out this method in view of the very interesting possibilities which it can offer in case of success. Of course, it is realized that unforeseen difficulties may still develop making the method unattractive.

References (van Eek)

Anon. 1976 Glomar Challenger sails on Leg 48. *Geotimes*, December, pp. 19–23.

Anon. 1977 World, Europe deep water-drilling records set. *Oil & Gas J.* 7 February, 34–35.

van Eek, W. H. 1977 *State of the art of environmental conservation, exploration and production.* United Nations Environment Programme – Industry Sector Seminars Petroleum Industry Meeting, Paris 29 March to 1 April. Published as *Petroleum Industry and the Environment Seminar Papers & Documents*, Vol. 1, pp. 59–100.

Goldman, E. C. 1978 Offshore subsea engineering. *Phil. Trans. R. Soc. Lond.* A **290**, 99–111. (This volume.)

Knebel, M. G. & Rodriguez-Eraso, G. 1956 The habitat of some oil. *Bull. Am. Ass. Petrol. Geol.* **40**, 547–561.

Koch, R. D. 1975 Drilling in deep waters: casing setting depths. *Colloque International sur les Techniques d'Exploration de Hydrocardures; Forage et Production en Mers Profondes, Paris.*

McLeod, W. D. R. 1976 A review of riserless drilling alternatives. *S.P.E. European Spring Meeting, Amsterdam.* Society of Petroleum Engineers A.I.M.E., S.P.E. 5768.

Meyer-Detring, D. 1976 Project of a floating drilling unit for waterdepths up to 2000 m. *Erdoel-Erdgas Zeitschrift,* International Edn 1976, volume 2, 33–36.

Peterson, M. N. A. 1975 Deep Sea Drilling Project technical contribution to deep offshore oil and gas exploration and eventual production. *Ninth World Petroleum Congress Proceedings*, vol. 4, PD(10)1, 3–14.

Watkins, L. W. & Howard, M. J. 1976 Buoyancy materials for offshore riser pipe. *Offshore Technology Conference:* O.T.C. 2654.

Phil. Trans. R. Soc. Lond. A. **290**, 125–133 (1978)   [125]

Printed in Great Britain

Manganese nodule mining

By K. B. Smale-Adams and G. O. Jackson

Rio Tinto Finance and Exploration Ltd, 6 St James's Square, London SW1Y 4LD, U.K.

This paper considers the mining of manganese nodules from the deep ocean at up to 5000 m, drawing attention to the essential need for a satisfactory legal régime under which mining companies can operate with security. The necessary exploration that has to be carried out before large investment can be made is indicated and the type and size of mining site required are determined.

The paper also considers the equipment required to collect, lift and transport the nodules. The most likely form of collection and lift is hydraulic but there are considered to be substantial development problems still to be solved. This indicates the need for large scale tests before the final decision on the mining system can be completed. The paper describes briefly a test collector operated in late 1974 and early 1975.

A brief description of the possible environmental problems is also included.

In considering the opportunities for commercial exploitation of manganese nodules one is indeed moving into deep water in all the senses that the phrase implies. This paper sets out to isolate and highlight those factors for consideration in planning a mining system. Most of the information is based on published information but we do in addition have the benefit of direct involvement with the Kennecott Copper Corporation. Clearly at this stage we are bound to some extent by confidentiality and commercial security and it is necessary to say that the views expressed are entirely personal and do not necessarily represent any corporate philosophy.

International régime

In speculating on the possibilities for economic exploitation one major difficulty is the lack of a legal régime. We feel it is necessary once again to repeat the urgency for some form of international legislation, or failing agreement at the current Law of the Sea Conference, national (reciprocal) legislation by those countries with the technological capability, to protect mining companies and to ensure the orderly development of this important new source of minerals for the benefit of mankind as a whole.

The régime should allow provisions for exclusivity and for security of tenure for the operator, adequate financial incentive to encourage development, bilateral or multilateral tax reciprocity, and at the same time a set of working regulations, environmental protection arrangements, and detailed financial provisions that will allow efficient utilization of the resource and a fair distribution of any rewards deriving from its exploitation.

Few operators would contemplate land-based operations without these fundamental provisions, and with the added risks to which a deep ocean operation would be exposed they are considered important and urgent. At the same time all mine operators recognize the status of the landlord or mineral rights owner – in this case 'the common heritage of mankind'.

The resource

In order to plan any mining operation one needs first to identify the orebody. Perhaps we should begin with the consideration of the total reserves available and the means of assessing the recoverable mineral content. There is little doubt that there is a potentially very large resource on the sea bed; there is also little doubt that given the opportunity, a number of potential operations will be offered for financing in the next 10–20 years.

There are no 'mine sites' yet identified as such, simply because none have yet been explored to the degree necessary for this purpose, at least as far as published information is concerned. Deep Sea Ventures have laid claim to one area of 60 000 km² between the Clarion and Clipperton Fault Zones and with three other consortia exploring in the Pacific it is likely that they too can identify 'areas of interest' leading to mine site development proposals. To put mine site identification into perspective, however, we have considered the occurrence of copper in Peru as an analogy in regard to mining development. Individual copper occurrences in that country number more than 500 and at first sight give an exaggerated idea of the potential reserves for exploitation. This, we hope, brings home the difference between an occurrence and a mine site. In a previous publication parameters necessary for a manganese nodule 'mine site' were identified as approximately an area of 30 000 km² with an abundance of 10 kg (dry)/m² at a grade of 1.3–1.4 % nickel, 1.1–1.2 % copper and 0.2–0.25 % cobalt.

Exploration and sampling

In order to assess the metal content of such a site with a confidence level of 90 % one requires at least 300 samples spaced on a lattice of approximately 10 km.

Samples are collected by free-fall samplers each comprising a set of clamshell type jaws mounted in a frame with concrete weights, a trigger, a float, a light and a transponder. These samplers are dropped overboard and sink to the bottom. When the trigger mechanism touches the sea bed the weights are released, the grabs close on the sample material and the float raises the sampler to the surface, where it can either be located by its light or from the transponder signal.

Additionally, box cores are used for lifting undisturbed *in situ* samples from the sea floor. It is possible to photograph the sample location on the sea floor just before the sample is collected.

Photography is used in its own right extensively for assessing nodule abundance on the sea floor, by counting individual nodules on each photograph. At least 30 000 photographs are available to us for this purpose. Television cameras equipped with videotape have also been of assistance as an additional means of assessing the abundance and the general nature of the bottom.

One also requires a topographical relief map to give assurance that any equipment on the sea bottom could be capable of negotiating the area to be traversed.

Sea mounts are not unknown on the ocean bed but generally speaking the sea bottom representation of a mine site has been compared to the rolling hills of Western Kansas, with slopes in excess of 6° rare (but present) and an occasional outcrop of soft volcanic ash or basalt protruding above the soft sediment. Since the occurrence of outcrops, especially basalt, could pose a potential hazard to mining equipment, an analysis of their frequency of occurrence and a means of avoiding and/or surviving collisions needs to be carried out as part of any determination of recoverable reserves.

Analysis of one area has indicated that these outcrops can range from a few feet to more than 20 ft and occur randomly at distances from a few hundred to over 20000 feet and appear to occur in clusters. Their occurrence is more frequent in areas of extreme slopes and large-scale features but they are also found in relatively flat areas. In any mine site it has been assumed that approximately 25–30 % of the area would be inaccessible to mining. The nodules lie in a single layer on a soft mud floor requiring very low bearing pressure on the contact surfaces of any equipment designed for nodule collection.

NAVIGATION

It has always been assumed if one is lucky enough to discover a mineral deposit in the deep ocean that one will be able to relocate it accurately and mine it in the exact configuration one wishes. Accuracy of location of samples and subsequent navigational control are therefore of paramount importance in planning the operation. On land one can be as accurate as the mining system allows, but in the deep oceans available long-range navigation systems mean that one may be up to 200 m from the point at which one aims. That point in turn can be 200 m away from where it has been plotted. This requires the establishment of a local navigational system with multi-ties for accuracy. The most accurate navigational aids are generally limited to a couple of hundred miles in range and are applicable to coastal waters. However, the longer-range low-accuracy systems, when integrated with a surface buoy system can be used in deep ocean navigation and within a given area accuracy is within 20 m. Having established a local grid the most efficient means of navigating beneath the surface is by the use of an acoustic beacon network where again accuracies of 15–20 m can be achieved.

A new development considered highly accurate is the Navstar Global Positioning System, consisting of 24 satellites in subsynchronous 12 h orbits. A minimum of four satellites will be in the range of any point on the globe at all times, allowing a three-dimensional fix worldwide. The system needs development, but it may substitute for the local system in the longer term.

OTHER MINE SITE CONSIDERATIONS

In our postulated mine site we have already assumed that 20–30 % of the area is inaccessible because of the topography. It would be reasonable to assume that parts of the area will also be of poor abundance and grade and one needs to consider for maximum recovery the need to avoid repeating traverses so that necessarily an unmined swathe is left between each run of the mining equipment. Once traversed by this equipment the surface-bearing pressure of the mud is lowered considerably. Nor is the mining equipment likely to pick up all of the nodules within the area of traverses. Combining these factors gives one a very low overall recovery efficiency within a mine site area and we have little reason to change the likely figure of 25 %. Improvement to a figure nearer 40 % can reasonably be expected in the second generation.

The theme we are trying to convey is that in spite of the 'enormous' overall reserves, the difficulties associated with its distribution, abundance, grade and value coupled with difficulty of access, extreme weather conditions and navigational surface difficulties are equally formidable. The lack of operating experience inhibits confidence but the prospect of challenge and subsequent reward are such that the four major mining consortia are interested in the possibilities for commercial exploitation.

Vast amounts of capital will be necessary and it is unlikely that mining companies will expend that capital other than with the support of their governments, their bankers and their technological advisers. It would seem a pity if the advance of science, or at least of technology, were to be inhibited by the law, or the lack of it, for peaceful purposes.

Having considered the nature of the resource, and its distribution, much ingenuity has been used to devise mining systems which will collect the nodules efficiently. The capital costs involved in mining at sea have such a high threshold value that it is imperative for economic reasons to mine on a large scale. Typically this would at 3–4 Mt/a. Allowing for down-time this is equivalent to 10–15 kt/day. The latter figure represents the expenditure of no less than 4 MW of power simply to raise the nodules against gravity from 5 km. Somewhat smaller mining systems can be considered if a market for manganese can be secured in addition to sales of nickel, copper and cobalt. A mining system for nodules is conventionally divided into four main components: (*a*) the collector, (*b*) the lift system, (*c*) the mining vessel, and (*d*) the transport vessels.

(*a*) *The collector*

Since the nodules lie in a single layer at the sea bottom, any collection device must sweep a large area per unit time to gather a larger quantity of nodules. Both mechanical (rakes, buckets, scoops, etc.) and hydraulic (suction dredge) types of collection have been considered. Because the sea bottom in prime nodule areas is composed of a soft, cohesive clay, mechanical devices, for example, are liable to fill with clay when dragged on the bottom, thus probably losing a good percentage of the nodules.

Hydraulic dredge systems work effectively in the clay bottom; however, the direct suction of the nodules results in a dilute slurry which can be expensive to lift unless means can be found to improve the slurry density during collection. Thus, slurry concentration is required on the collector prior to lift pumping.

Both self-propelled and towed collectors have been considered. The self-propelled collector has the advantage of having some directional and speed control, but has been found to present several difficult design problems:

(i) Traction on a clay bottom is difficult to achieve, particularly bearing in mind the very low bearing pressure of the mud floor.

(ii) With self-propulsion the lift pipe is nearly vertical, requiring a heavy collector and high power on the collector. Furthermore, small depth changes require substantial alteration of lift pipe scope.

In one version of a collector, the manganese nodules are collected by hydraulic suction, possibly with the aid of mechanical or hydraulic nodule-dislodging devices such as tines or water jets. The feasibility of the hydraulic suction process has been well established by sea trials and other land based tests. The quantity of nodules lifted by such a typical collector depends on the average nodule abundance, the width and forward speed of the collector, and the efficiency with which the collector works. Efficiency can be decreased not only simply because nodules fail to be raised, but because it is difficult to ensure that the whole frontal area of the collector is full utilized. With these limitations in mind a typical collector would be 10–15 m in width.

Nodules, sediment, other solids and water are sucked into a collector by the flow generated by a pump located near the collector head. It will probably be necessary to provide some means of concentrating the sediment–water mixture, perhaps by taking advantage of the significant difference in settling velocities between diluted sediment fines and the nodules. There is, however, an unavoidable loss of nodules at this point. The solid particle concentration of the slurry in the lift pipe is likely to be between 10 and 15 % by volume. It is possible that heavier solids (such as large pieces of basalt) can be removed at this stage and do not therefore pass into the feed pipe.

It is possible that some obstacle detection capability needs to be incorporated into the design of the collector and that short-term avoidance may need to be carried out in order to avoid obstacles not previously identified by either the mining vessel or the mine pre-survey. Such encounters will, however, shorten the availability of the collector for normal mining operations. The maximum height and shape of solid obstacles that can be accommodated by the collector is an important design variable of the collector and normally has repercussions both on the detail necessary on the mine survey plan and the steering control necessary on the sea bottom.

The collector needs up to several hundred kilowatts of electric power to operate the suction pumps, metering devices and instrumentation. Power and instrument cables and connections have to be designed to withstand the enormous pressure at 5 km. Submerged electric motors have also presented a challenge to technology but this problem has been overcome on a test scale.

The instruments and sensors in the collector need to carry out the following functions: obstacle detection, communications, performance evaluation, and operation monitoring.

A successful demonstration of a hydraulic suction collection device was carried out in late 1974 and early 1975 at approximately 5000 m. The scaled-down model used in this test was designed to be towed over the ocean floor at depths ranging from 4 to 6 km by a special, steel-armoured, electromechanical power cable. The vehicle tow velocity was approximately 2 knots (1 m/s) with an expected speed range of 3 m/s. The front section of the vehicle was designed to accommodate obstacles and surface irregularities and to be capable of surviving direct frontal vertical wall-type impacts without structural damage. In case the vehicle capsized, a design was developed which would permit it to roll back into the normal tow position.

The test was operated at a collection rate of 2 kt/day in a real mine situation. This success has given encouragement to proceed to the development of integrated tests of both collection and lifting nodules.

Instrumentation was fitted to detect, measure and record the following:
(a) nodule mass flow through the duct system;
(b) duct flow velocity;
(c) vehicle forward velocity and acceleration;
(d) vehicle heave displacement;
(e) vehicle pitch displacement;
(f) collector head heave displacement;
(g) relative position of vehicle to bottom;
(h) vehicle depth;
(i) tow force at vehicle tow point.
In addition, a forward-looking camera was built into the nose section of the vehicle.

(b) *The lift system*

Other than the collector, the lift represents the most critical element of the mining system. It must be designed to carry the high loads imposed by its own weight and the hydrodynamic towing forces. Various methods for lifting nodules have been considered, both mechanical and hydraulic. These have included the following:

(i) mechanical lift:

continuous line bucket (c.l.b.);

dragline bucket;

batch lift.

(ii) hydraulic lift:

airlift hydrocarbon or light solid particle lift;

two pipe (with slurry injection at the collector);

in-line centrifugal pumps, electrically driven;

in-line centrifugal pumps, shaft driven;

in-line mixed flow pumps, shaft driven;

in-line axial flow pumps, shaft driven;

in-line axial flow pumps, water turbine driven.

In general, mechanical lift systems depend on the strength of synthetic rope and the speed of traction equipment to achieve a high lift capacity. The drag-line bucket method is obviously limited by the round trip time and the size of a single bucket. The continuous-line bucket method has the advantage of running a continuous stream of buckets but is limited by the filling efficiency of the buckets, the rope strength and speed of operations. Studies have indicated that the limit set to the nodule recovery rate by synthetic ropes and the traction machinery is roughly 3 kt/day. The practical limit is probably less than this.

A batch lift method has already been proposed which does not have the limitations of rope strength of traction speed of the c.l.b. In this concept, a cable linking the mining ship to the collector simply acts as a guide for a large container which shuttles back and forth between the collector and the ship. This container is filled with nodules at the collector by means of a conveyor. Power to raise the shuttle container comes from buoyancy obtained by pumping ballast tanks empty after nodules are loaded. The container is hydrodynamically streamlined so that it accelerates to a high speed before finally docking with the ship where it is unloaded. This process was shown to have a theoretically unlimited capacity although it embodied numerous complex mechanical devices which could lead to unreliable operation.

All the hydraulic lift methods may be thought of as containing two sections; a pipe section and a pump section.

The lower two-thirds of the lift system in nearly all hydraulic methods consists of a single large diameter pipe. The exact diameter of the pipe is determined by optimizing such factors as head loss, nodule size and fabrication capabilities. For large production systems, the head losses in the lift pipe can be in excess of 10 MPa. Thus the first stage of the lift pump must be submerged over 1 km to generate the required suction without cavitating.

The differences between the various hydraulic lift methods arise from the ways of achieving this suction force in the upper section of the lift system.

Two methods, the airlift and the light solid particle lift, are designed to accomplish this by decreasing the average density of the fluid in the upper sections in order to generate upward flow. Using hydrocarbon for the same purpose raises objections on environmental pollution grounds. The airlift method operates by injecting compressed air at a depth of approximately 1–1.5 km. The hydrocarbon lift injects a light liquid instead. Both methods have the advantage of requiring no moving parts below the waterline. The airlift has several disadvantages, however:

(i) lift capacity is limited by the volume of the lift pipe which can be allowed to be occupied by air in the upper sections without causing solids to settle;

(ii) variations in load can cause system shutdown;

(iii) power requirements are higher than for an equivalent mechanical pump.

While the hydrocarbon and light solid particle lift does not present the problems encountered because of the compression and expansion of air, the density differences between available hydrocarbons or solid particles and water are so slight compared with air and water that capacities are severely limited.

Mechanical pumping methods appear to provide a feasible means for lifting nodules provided sufficient development work is carried out. The power required to raise the nodules using either airlift or mechanical pumps turns out to be of the same order and around 12–15 MW. The attractive simplicity of the airlift system, which avoids all undersea moving parts, does not require any sacrifice of power. However, no airlift system comparable in size to that required for nodules is operated anywhere in the world and there are fundamental design problems yet to be resolved.

The emergence in recent years of high-powered submerged electric motors has made it feasible to consider them as the power unit for nodule pumps. Nevertheless, there is much development work to be done to extend their power range before such pumps can be said to be proved. There does not appear to be unanimity among the various consortia on the type of pumping system that should be used.

(c) The mining vessel

The mining vessel is designed to operate in the equatorial Pacific and remain at sea for periods of up to 4 years without returning to shore facilities. The general operating ground rules dictate that all victuals, fuel, stores, repair parts and personnel be transported to the mine site and transferred to the mining ship. All logistics supplies and personnel will be transported by means of the ocean transport vessels. All normal maintenance on the mining ship needs to be carried out at sea except such maintenance that would require docking or extensive tear-down. Docking for this purpose is envisaged as occurring approximately every 4 years. The crew size is expected to be between 100 and 120 individuals, including mining staff, pipe handling crew and the ship's operating crews. Crew change is expected to occur every 30 days. Experience of this type of operation is available to oil companies in their off-shore exploration programmes.

The gimbal for supporting the lift pipe needs to be considered in some detail. A roller-bearing gimbal of the required load-bearing capacity is understood to have been used in the *Hughes Glomar Explorer*. The heavy tow load requirements are unique, however, and this has resulted in the requirement that the roller bearings of the pitch axis be able to absorb axial thrust as well as radial loads.

The use of a flexure gimbal to support 35 MN is beyond what has been undertaken in the past by several orders of magnitude.

9-2

The determination of scantlings and the midship section by normal design rules is inadequate for a mining-ship design. A detailed structural design needs to be developed and a finite element analysis undertaken before the mining vessel arrangement is agreed upon.

Because of the large drag generated by the lift pipe the power requirement to drive the ship forward at up to 2 m/s can be as large as 22.5 MW. In addition, the pipe will shed eddies alternately on each side and cause the whole pipe to vibrate at a frequency of the order of 1 Hz. This can do considerable damage both to the pipe and to the ship. For both drag and oscillation reasons, fairings have to be introduced which need to be attached to the pipe as it is deployed. Even so, the towing forces are an order of magnitude larger than any ocean-going tug can provide. The large amounts of power (40–50 MW) required on a continuous basis are ideally suited to a nuclear power source and this may well be the choice in the long run. However, the construction time allied to the long-winded processes of licensing makes oil firing the only real choice for the early mining systems.

The manoeuvring capabilities of the mining vessel need to be integrated into a mining system study.

The pipe handling system dynamic analyses which have been performed have indicated that it is feasible to deploy and retrieve the pipe string with the collector attached in up to a sea state 4 at the maximum speed of 7 m/min and to tow the pipe string and collector at speeds between 1.5 and 2 m/s with fairings. The collector and pipe string can be deployed and recovered without heave-motion compensation on the pipe handling system. The pipe string must be isolated from angular roll-and-pitch ship motions in excess of 1° at all stages of deployment. This is in order to prevent excessive bending stresses from occurring in the pipe at the support point. The maximum pipe string drag angle occurring during mining operations is likely to be less than 25°.

(d) Ocean transport

The nodules, once delivered on to the mining vessel in a dilute slurry, need to be concentrated still further in order to reduce the transport costs. This concentration can be carried out either on the mining vessel or on the transport. If it is carried out on the mining vessel, which has the attraction that the concentrating equipment does not need to be duplicated on each transport, it becomes necessary to transfer solid or semi-solid material from the mining vessel to the transport vessel. If, on the other hand, slurry is concentrated after transfer it can be transferred as a slurry in a hose. In either case strict station-keeping requirements are imposed on both vessels and particularly so in the case of the solids transport. The problem is one of maintaining station for days while travelling at a relatively low speed of a few knots in a situation where sea states may be changing and where the mining vessel has to carry out a strictly predetermined set of manoeuvres. The vessels of course are themselves substantial; the mining vessel being perhaps 45 000 tonnes displacement and the transport vessel of 65 000 tonnes dead weight. Three or four transport vessels would maintain a continuous round trip from mining vessel to shore station. The mining vessel would contain a small buffer store of nodules in order to cope with the changeover from one transport vessel to another or to help in particularly awkward manoeuvring positions.

The environment

It is difficult to speculate on the effect on the environment of deep sea mining, but large-scale test units which are likely to be used in the next stage of commercial development will enable strict monitoring of the likely disturbances to organisms in the benthic zone to be carried out, and the effect of resettlement of disturbed sediments on bottom organisms to be examined. It will also be possible to assess the effect which the plume around the mining vessel has on light penetration of near-surface waters and of the increase in dissolved nutrients in the euphotic zone.

The deep ocean environmental studies, however, suggest that potential pollution will be minimal.

Discussion

B. White (*Department of Mineral Resources Engineering, Royal School of Mines, Prince Consort Road, London, S.W.*7). I should like to ask about the methods used to evaluate the resources.

The box corer described collects samples of the nodules and can be likened to the diamond drill used to sample conventional ore-bodies which yields a core for assay. We are informed by most authors that the relative abundance and the metal content of the nodules is very variable over comparatively short distances.

If it is not possible to control the location of the sampling positions due to the effects of currents on these free-fall devices, what statistical techniques are used for the valuation of the resource?

K. B. Smale-Adams. I agree that *precise* control of the sampler is not possible; the deposits are not too variable within localized areas but in no areas have sufficient samples been collected to justify a high level of statistical confidence. Within any 'potential mine site areas' considerably more sampling would need to be undertaken to establish an acceptable confidence level.

Phil. Trans. R. Soc. Lond. A. **290**, 135–152 (1978)   [135]
Printed in Great Britain

Engineering aspects of manned and remotely controlled vehicles

By R. F. Busby

R. Frank Busby Associates, 576 South 23rd Street, Arlington, Virginia 22202, U.S.A.

Mushrooming activities in offshore oil and gas developments have produced a wide variety of manned and remotely controlled vehicles which are conducting many tasks traditionally performed by surface craft and/or ambient-pressure divers. Trends in present underwater vehicle design and work requirements of both vehicles and divers indicate that direct and remote viewing, manipulative dexterity equal to the diver, and diver lockout support are deep-water work requirements. Diver lockout submersibles capable of operating to 2000 m are technically feasible, but saturation decompression schedules at this depth are not foreseen within the next decade. Substitution of mechanical means for human capabilities to perform diver-equivalent work will require major improvements and technological break-throughs in the areas of manipulation, wireless signal transmission and power sources. Individual or combined application of manned and remotely controlled vehicles offer the most immediate solution, but environmental factors and technical deficiencies combine to reduce their effectiveness. Design of future undersea hardware for manipulation by mechanical devices and inspection/testing by mechanical means can significantly narrow the performance gap between human and mechanical devices.

Introduction

Deep water is no stranger to the ocean engineer. The technical problems involved in reaching and operating at 2000 m with manned and remotely controlled vehicles were met and surmounted in the mid-1960s. New materials, life support systems and components for deep submergence have been continually developed since Trieste I reached 10911 m in 1960. From the point of view of technical feasibility, reaching and operating at 2000 m depth presents no problems. However, transferring the full range of 300 m work capabilities now available to a depth of 2000 m and, by necessity, replacing the manipulative dexterity, responsiveness and agility of the ambient-pressure diver with a manned or remotely controlled vehicle does present a wide range of engineering problems. The potential solutions are not only hardware-orientated, but they involve design philosophy and employment techniques as well.

The substance of this paper is to predict or anticipate engineering problems likely to be encountered by manned submersibles and remotely controlled vehicles at 2000 m depth. It is therefore appropriate to identify first what is now available in these vehicles, what tasks they perform in offshore oil, and what trends are seen in design and capabilities. Further – and this is a perilous undertaking – the likelihood of support from the ambient pressure diver at 2000 m must be predicted, for if he cannot be employed, then the potential problems increase by more than an order of magnitude.

Vehicle status

Manned submersibles

In 1970 the manned submersible was fast becoming an endangered species. With the advent of North Sea oil and gas discoveries its numbers multiplied. For comparative purposes table 1

TABLE 1. SUBMERSIBLE CONSTRUCTION AND BUYERS/USERS 1970–1976

year	submersible	buyers/users
1970	Cyana	government (civil)
	Nekton Beta	industrial
	PC-9	industrial
	SDL-1†	government (military)
1971	Burkholder	industrial
	Hakuyo	industrial
	Nekton Gamma	industrial
	PC8B	industrial
	Sea Otter	industrial
	Johnson-Sea-Link I†	research
1972	Mermaid II	industrial
	Pisces IV	government (civil)
	PS-2	industrial
	Globule	industrial
1973	Griffon	government (military)
	Pisces V	industrial
	Sea Ranger	industrial
	Vol-L1†	industrial
	Skadoc 1000†	industrial
1974	Diaphus	academic
	Aquarius I	industrial
	Moana I	industrial
	Johnson-Sea-Link II†	research
1975	PC-1201	industrial
	PC-1202†	industrial
	PC-14C-2	government (military)
	Argus	government (civil)
	Pisces VII, XI	government (civil)
	Pisces VIII, X	industrial
	Mermaid III†	industrial
1976	Leo	industrial
	Moana III, IV, V	industrial
	PC-1203	industrial
	PC-1204	industrial
	Vol-L2, L3†	industrial
	PC-1801†	industrial
	PC-1802†	industrial
	PC-16†	industrial
	Taurus†	industrial
	Mermaid IV†	industrial
	PRV-2†	industrial
	URF†	government (military)

† Diver lockout capability.

is included, which shows the growth in vehicle production during the past few years. The major customer for submersibles today is industry, and the major user of industrially owned submersibles is the offshore oil and gas industry. An inventory in 1976 of worldwide manned submersibles showed that there were a total of 91 vehicles; their status was as follows: operational/sea trials, 57; under construction, 16; undergoing refit, 7; inactive, 11. Because the field is dynamic, these values can change quite rapidly. Not included are perhaps 15–20 shallow-diving, one-man vehicles built for recreational use.

It is very difficult to generalize when discussing design and capabilities of manned submersibles. Only a handful are identical and even within these there are variations. However, to gain an appreciation of the industrial field at large the following characteristics are given:

(*a*) The average maximum operating depth is 572 m; the deepest is 3000 m (the French submersible *Cyana*); the average length, beam and height are 6.2, 2.3 and 2.7 m respectively.

(*b*) All use lead acid batteries.

(*c*) Crew complement is from two to six.

(*d*) Dive working duration is from 6–8 h.

(*e*) The average cruise speed and endurance is 1 knot for 7.9 h.

(*f*) The average payload is 480 kg.

(*g*) Dry mass is from 2–26 t.

(*h*) About half of the newly constructed vehicles have diver lockout capability.

(*i*) Approximately 80 % carry at least one manipulator; 40 % of these carry two.

(*j*) Launch/retrieval can be generally conducted in sea-state 4 and, in some instances, sea-state 7.

The major exception to the above is the 'Auguste Piccard'. Being 29 m in length and 168 t in mass and having a life-support duration of 90 man-days, it is in a class by itself.

Navigation or positioning capability of submersibles varies from company to company, but position accuracies of ± 1 m within an area of 130 km² are attainable relative to bottom-mounted transponders.

Manoeuvring characteristics vary widely but thrust, yaw, heave and pitch control are general capabilities. Mid-water hovering is also common, but to stabilize the vehicle when working on a fixed structure it is a general practice to grasp the structure with one manipulator and work with the other.

Work tools – e.g. drills, wrenches, grinders, brushes, etc. – are available to varying degrees on all vehicles. The most dominant work capability is direct viewing coupled with t.v. video documentation.

Unlike many other industries, the major submersible builders do not produce a fleet of similar vehicles on the speculation that buyers will be found. Each vehicle is generally built under contract, and each one is somewhat different from its predecessor. The difference might be in depth, lockout or non-lockout capability, size, instrumentation and crew. The result is that each vehicle reflects the buyer's idea of present and future capabilities required to meet the needs of the offshore customer (i.e. offshore oil and gas). Consequently there is little likelihood that a fleet of obsolescent vehicles will exist in the near future, such as the next five years.

A further consequence of this one- or two-at-a-time purchasing is that the size of the submersible fleet keeps pace with the demand for vehicle services. No operating company intentionally orders more vehicles than it can see a need for, and all operators are keenly aware of offshore activities that may provide a market for their services. The present situation therefore is one where vehicle supply and demand is equal, and will probably remain so unless there is a major change in underwater work requirements.

Remotely controlled vehicles

There are several types of vehicles which fall into this category: tethered, free-swimming vehicles; tethered, bottom-crawling vehicles; towed vehicles; and untethered, free-swimming vehicles. This discussion is limited to the tethered, free-swimming vehicles of the RCV-225 variety.

Undoubtedly the most dynamic growth in a particular underwater platform has been exhibited by the remotely controlled vehicles (herein they will be called RCVs; RCV is a registered trademark of Hydro Products, San Diego, CA.). In 1974 there were approximately eight RCVs; today there are at least 40. A listing of these vehicles and their depth capability is contained in table 2.

TABLE 2. UNMANNED, SELF-PROPELLED, TETHERED VEHICLES

vehicle	depth/m	builder
Angus	300	Heriot-Watt University, Edinburgh, U.K.
Angus 002	300	Heriot-Watt University, Edinburgh, U.K.
Consub 1	610	British Aircraft Corp. Ltd Bristol, U.K.
Consub 2	610	British Aircraft Corp. Ltd Bristol, U.K.
Cord	457	Harbor Branch Foundation Ft Pierce, Fla., U.S.A.
Curv II	762	Naval Undersea Center San Diego, Calif., U.S.A.
Curv II	762	Naval Undersea Center San Diego, Calif., U.S.A.
Curv III	3048	Naval Undersea Center San Diego, Calif., U.S.A.
Cutlet	305	Admiralty Underwater Weapons Establishment, Portland, U.K.
Deep Drone	610	Supervisor of Salvage Washington, D.C., U.S.A.
Eric	500	French Navy Toulon, France
Eye Robot	100	Mitsui Ocean Development & Engineering Co., Ltd Tokyo, Japan
Manta 1.5	1500	Institute of Oceanology Moscow, U.S.S.R.
RCV-150†	1829	Hydro Products San Diego, Calif., U.S.A.
RCV-225‡	2012	Hydro Products San Diego, Calif., U.S.A.
Recon II	457	Perry Ocean Group Riviera Beach, Fla., U.S.A.
Ruws	6096	Naval Undersea Center Honolulu, Hawaii, U.S.A.
Scarab I, II	1829	Ametek Straza El Cajon, Calif., U.S.A.
Sea Surveyor	220	Rebikoff Underwater Prod. Ft Lauderdale, Fla., U.S.A.
Snoopy	457	Naval Undersea Center San Diego, Calif., U.S.A.
Snoopy	457	Naval Undersea Center San Diego, Calif., U.S.A.
Snurre	600	Royal Norwegian Council for Scientific Research Oslo, Norway
Telenaute	1000	Institut Francais du Petrole Paris, France
Trov	366	McElhanney Offshore Survey & Engineering, Ltd Vancouver, B.C.
Trov OI§	366	McElhanney Offshore Survey & Engineering, Ltd Vancouver, B.C.

† Three vehicles total: Martech International, Houston, Tx., U.S.A., Scandive, Stavanger, Norway; Deep Sea Resource Dev. Corp. Taiwan, Formosa.

‡ Eight vehicles total: Seaway Diving, Bergen, Norway (2 vehicles); Martech International, Houston, Tx., U.S.A. (2 vehicles); Sesam, Paris, France (2 vehicles); Taylor Diving & Salvage, Belle Chasse, La., U.S.A. (1 vehicle); Esso Australia, Ltd, Sale, Australia (1 vehicle).

§ Two vehicles total: Underground Location Services, Glasgow, U.K.; British Petroleum, Middlesex, U.K.

RCVs are as varied in design as are manned vehicles, and generalities regarding their characteristics are attended by numerous exceptions.

The basic tethered, self-propelled vehicle system consists of the vehicle itself (and sometimes an underwater clump or launcher), a cable and a shipboard control/display console. Supporting equipment includes a launch/retrieval device, a cable winch, an enclosed area for the vehicle operators and shipboard components and, if shipboard power is not available or suitable, a power supply unit.

Vehicles owned by industrial users range in depth capability from 200 m to 2000 m; the average is 1300 m. Depth *per se* presents no problem to the RCVs. Control of the vehicles at great depths is a problem which is discussed later.

Most vehicles are constructed of an open metal framework that supports and encloses (for protection) its various components. Buoyancy is generally positive by a few kilograms when the vehicle is submerged; this provides a fail-safe assurance that the vehicle will surface in the event of a power failure. Generally, but not always, syntactic foam blocks mounted atop the framework provide the required buoyancy.

TABLE 3. WORK INSTRUMENTS

	viewing/photography				manipu-lator	sonar		current meter	ther-mistor
	t.v.	still	stereo	cine		search	homing		
Angus	×	·	·	×	·	·	×	·	·
Consub 1	×	·	×	·	·	·	·	·	·
Consub 2	×	·	×	·	·	·	·	·	·
Cord	×	·	·	·	×	×	·	×	×
Curv II	×	×	·	·	×	×	×	·	·
Curv III	×	×	·	·	×	×	×	·	·
Deep Drone	×	×	·	·	·	×	×	·	·
Eric	×	×	·	·	×	·	·	·	·
Manta 1.5	×	·	·	·	×	·	·	·	·
RCV-225	×	·	·	·	·	·	·	·	·
RCV-150	×	·	·	·	×	·	·	·	·
Recon II	×	·	·	·	×	·	·	×	·
Ruws	×	·	·	·	·	×	×	·	·
Scarab I and II	×	×	·	·	× (2)	·	·	·	·
Sea Suveyor	×	·	·	·	·	·	·	·	·
Snoopy	×	·	·	×	·	·	·	·	·
Telenaute	×	·	·	×	×	·	·	·	·
Trov	×	·	·	·	× (2)	·	·	·	·

+ lights on all RCV's

The underwater component(s) or 'vehicle' of these systems weigh from 68 kg to as much as 2268 kg. The sea-state limitations on launch/retrieval are controlled by the nature and sophistication of the shipboard handling equipment. Some indication of sea-state limits can be gained from the following operator statements: Consub 1 can be launched/retrived up to sea-state 4; Deep Drone is designed to be handled up to sea-state 5 if 'normal' handling equipment is available which is generally employed to handle manned submersibles. These two are not the heaviest vehicles operating, but they do fall around the average vehicle mass of 961 kg.

The speed of RCVs is similar to that achieved by manned vehicles, and ranges, at the surface, from 1 to 5 knots (1.8–9.3 km/h). There is a decrease in speed with depth and/or with increase in currents which may range from 20 to 84 % of the surface speed. The reduction is caused mainly by cable drag, but can be alleviated by different modes of vehicle deployment. The Scarab vehicles are designed to cruise along the bottom while (in conjunction with the surface ship) they tow the entire length of cable. The RCV-225 is deployed from a launching cage and works around the launcher on 120 m of tether cable; hence, cable drag is substantially reduced. For this reason, many of the RCVs employ a launcher or clump.

All but a few vehicles are capable of two translation motions and one rotational motion; these are thrust (forward/reverse) and heave (up/down) and yaw (left/right heading changes) respectively. These motions are provided by the arrangement of two horizontal or forward thrusters and one vertical thruster. By adding a fourth lateral or side thruster a third translational motion is obtained: sway or sidle. If the lateral thruster is mounted forward, it is used to augment yawing, rather than providing a sideways translational motion.

For routine operations the support crew complement ranges from one to seven; three to four is average.

The instruments listed in table 3 are those which are standard onboard equipment. All RCVs carry underwater lights. British Aircraft Corporation's Consub 1 has a manipulator-held rock drill which has successfully operated in the field, but is not listed in table 3. The majority of RCV manipulators are simple devices which can do no more than extend and open/rotate the claw. The limited orientator and locator motions are not a liability because the vehicles themselves can provide several more degrees of freedom to the manipulator by virtue of their excellent manoeuvring capability.

Navigation or positioning is similar to that used on manned vehicles with variations in capabilities from company to company.

Also like their manned counterparts, RCVs are generally built to order, but the variation in design or capability between vehicles of a particular series is slight. Payload, or the ability to carry additional submerged mass (i.e. work tools), is very small and without modifications, generally limited to no more than 1 or 2 % of the vehicle's mass.

WORK TASKS AND TRENDS

Submersible and RCV support for offshore oil and gas is arbitrarily divided into two categories: (1) tasks historically performed by divers, and (2) the provision of observations and measurements of the bottom or hardware which require details that conventional over-the-side surface techniques cannot attain. The categories are further divided into three functional tasks: (i) observational/documentation; (ii) observational/manipulative and (iii) observational/measurement/sampling. The 'observational' function is included in each task to emphasize that without some degree of visibility, the tasks now performed by undersea vehicles could not be conducted. The need for visibility is significant because, as is discussed later, the ambient diver performs a great deal of work by feel on objects he cannot see clearly.

Following are various tasks which have been performed by manned submersibles and RCVs for offshore oil and gas.

submersibles	RCVs
Observational/documentation tasks	
pipeline route inspection	geological observations
pipeline arc observation	cathodic protection inspection
video/visual survey of installed pipelines	pipeline inspection
cathodic protection inspection	tank and buoy inspection
platform site inspection	bottom reconnaissance
platform jacket inspection	marine fouling assessment
pipeline inspection	
pipeline burial inspection	platform inspection (pre- and post-installation)
inspect/videotape manifold and hose of spar buoy	subsea completion system inspection
inspect/videotape legs, bracings, members, anodes and scour on drill platform	anchor dragging assessment
inspect/videotape chains, ancisors, anodes and transponders of drill platform; assist in change of hose string	wreck identification
pre-burial pipeline inspection	
inspect/videotape loading platform	
locate/inspect/videotape pipeline	

Observational/manipulative tasks

object removal prior to pipeline trenching

loosen and tighten bolts with impact wrench

drill holes in steel structures

collect hard rock core samples

stud insertion (explosive embedment)

close/open pipeline valve handwheel

wire brushing for inspection and maintenance

torch burning (acetylene) and concrete
 chipping

cable inspection and burial

assist in trenching operation

inspect and help disconnect and hook up an
 experimental oil storage tank

cable and repeater burial

drillship guideline change out

preparation of an abandoned wellhead for
 re-entry (template alignment; guideposts;
 install new guidelines)

maintenance of buoy moorings and offshore
 platforms

small object retrieval

rock fragment collection

hard rock drilling

benthic organism collection

drill bit recovery

assist in connecting surface retrieval line

Observational/measurement/sampling tasks

submersibles

pipeline route survey and sampling (side scan sonar, echo-sounder, rock and sediment sampler)

platform site surveys

establish/document/measure length and height of suspended pipeline sections

post-pipeline entrenchment profile (echo-sounder)

This is not an all-inclusive list of the tasks manned submersibles and RCVs are now conducting in the offshore fields, but it is a representative sample. The last category is a relatively new, but promising, rôle, i.e. acoustic mapping in conjunction with sampling and observations. Various attempts were made in the middle and late 1960s to use submersibles as undersea mapping platforms, but these were experimental exercises. With increasing depth, details of bottom topography are more difficult to obtain from a heaving, pitching vessel than from a stable, submerged vehicle. This task can be expected to become more frequent as production proceeds into rough, deep areas. There are no reported efforts whereby RCVs have been used as acoustic mapping platforms; possibly the present lack of payload to accommodate required instrumentation is at fault. The British Aircraft Corporation, however, does include in its new Consub design the capability to exercise such options. In spite of the dynamic growth of RCVs, they are still feeling their way and have yet to realize their full potential.

Trends

Manned submersibles

If consideration in design and capabilities is only given to submersibles currently used by industrial firms, then the following trends can be seen: diver lockout and dry transfer is an increasing capability (8 out of 14 vehicles built in 1976 offered diver lockout); plastic bow domes for increased viewing are mandatory; greater electrical (i.e. battery) power is sought and the dry mass of vehicles has increased (the newly built Taurus weighs 26 t). In short, industrial submersibles are now larger, more powerful and offer a wider range of instrumentation and diver support than did the vehicles of the 1960s and early 1970s. The weight increase is a reflexion of greater depth (stronger materials) and battery capacity. Less obvious is the shape of the pressure hull. In vehicles operating in depths greater than 600 m the pressure hull is spherical because it is the most efficient shape for dealing with pressures below this depth. A sphere, however, is the least efficient shape for capitalizing on layout within a particular volume: the cylinder is the most effective for this.

A very recent trend in submersible design is Oceaneering International's Arms (Atmospheric Roving Manipulator System) and the two subsea completion diving bells of Comex. Both systems are manoeuvrable spheres connected by a lift/signal transmission cable to the surface and equipped with sophisticated manipulators. Significantly, these systems were designed for employment from specific drillships and for conducting inspection and manipulative tasks. Guide rails affixed to the blowout preventor stack allows the Arms to be secured to a rail and then move circumferentially around the stack by means of a built-in friction drive system.

Both systems depart from the typical manned submersible in that they have forsaken extended bottom cruising for precision manoeuvring and control. Arms, for example, is designed to maintain position within 1 m of any part on the outside surface of the blowout preventer. A further emphasis has been placed on manipulative dexterity. The Arms manipulator is the most sophisticated in that if offers a force-feedback capability which is intended to allow the operator to 'feel' the task being performed. Both systems were built in late 1976 and scheduled for operations in the spring of 1977.

RCVs

The industrial operational life of these vehicles has been so short that major trends are hard to discern. Indeed, in many instances the techniques of employing these vehicles are still in what might be termed the development stage. As with any new capability, there is a period of trial and error to see where the vehicle offers its best application. It has become obvious that the RCV is an excellent viewing and t.v. documentation platform and can be used to perform straightforward manipulative tasks, but under what environmental conditions and with what surface support have not been precisely defined.

The most obvious trend to date has been the incorporation of a launcher or clump into the umbilical cable. The launcher acts to keep the power/signal transmission cable taut and absorb the effects of heave imparted by the surface ship. With this arrangement the RCV is able to work, unhindered by surface motion, from a tether cable attached to the launcher.

A more recent trend is toward larger vehicles with greater payload capacity. The RCV-225 weighs 80 kg; the evolutionary extension of this vehicle, RCV-150, weighs 220 kg and is capable of accommodating a wide variety of instrument options. A similar trend is seen in follow-on vehicles to Consub 1. Additional equipment capabilities to the new vehicles are search sonar, vehicle tracking devices and manipulators. The RCV-150 also includes an option for head-mounted display and control. Much like the early proponents of manned submersibles, the proponents of RCVs now speak of multi-purpose vehicles equipped with a wide array of devices for surveying, sampling, inspection and manipulative work tasks (e.g. brushing, drilling, grinding, etc.). The attainment of such versatility is more complex, however, than merely attaching another instrument.

Paralleling the specialized design of the Arms and Comex diving bells are two RCVs designed by Hydrotech International and Exxon.

The Hydrotech system is still in the design state and is scheduled for completion by 1979. The system is designated as an unmanned deep water (1200 m) pipeline repair system, and consists of two vehicles: a work vehicle and a vertical transport vehicle. The work vehicle is designed for soil excavation, pipe coating removal, pipe cutting and end preparation; its dry mass is 49 t. The vertical transport vehicle is designed to transport and position replacement sections; its dry mass is 63 t. Both vehicles are cable-connected to the surface and both rely upon closed-circuit t.v. for real-time information. Since underwater visibility is mandatory in order to conduct work, a clear-water flushing capability is included on each vehicle. Model tests indicate that launch/retrieval is possible in sea-state 3.

The Exxon system has undergone limited field testing and is scheduled for near-future work in the Gulf of Mexico. The device is termed a Maintenance Manipulator System (M.M.S.) and is designed to perform routine maintenance on Exxon's Submerged Production System (S.P.S.). Failure mode prediction of S.P.S. components identified those likely to malfunction with wear or by external damage. These components were then designed for removal, replacement and pressure testing by the M.M.S. The M.M.S. is guided to the S.P.S. platform along a pop-up buoy line; it then mates with a cogged track on the platform which is routed to place the M.M.S. in position to work on the pre-isolated, faulty component. When replacement is completed, the M.M.S. transports the faulty component to the surface where the system is retrieved. Monitoring of the work is by closed-circuit t.v. Underwater visibility is therefore critical.

TABLE 4. AMBIENT DIVER WORK TASKS

welding	.	rigging	×
drilling	×	bolting/unbolting	×
cutting	×	assembling	.
grinding	×	grouting	.
inspection (visual)	×	painting	.
measurements (dimensional)	×	site investigation	×
testing (non-destructive)	×	directing surface lifting/lowering	×
video documentation	×		

THE AMBIENT PRESSURE DIVER

The variety and difficulty of potential problems encountered in working to 2 km by manned and remotely controlled vehicles will be determined by the depth to which the ambient diver can work. If the diver cannot support offshore oil to 2 km depth, then the problems will be of great magnitude. At present both vehicles are competing in many tasks with the diver. In several applications they are as effective and in some they are better. But in other tasks they cannot even approach the diver's performance.

Offshore oil diving can be, for convenience, placed in two categories: scheduled and non-scheduled. Scheduled diving is that where the diver's rôle is known before the dive. When, for example, a structure is planted on the seabed the diver's rôle is predetermined and he may have trained extensively to perform it. Non-scheduled diving is that where the diver responds to an accident or malfunction, for example, recovering a lost tool, repairing a broken structure, or tightening a loose nut. In both cases the job the diver performs may be similar (e.g. welding, cutting or rigging) but in the first instance the conditions under which it will be performed are controlled; in the second they are not. The difference is critical.

Table 4 is a general tabulation of the types of work performed by the offshore ambient diver. The × to the right indicates tasks which manned and remotely controlled vehicles have also performed.

There are several aspects of these tasks which are significant, the most important being that the diver can do these virtually anywhere on a structure, the vehicles are restricted by virtue of size or by their umbilical to working on the extremities. A second important aspect is that the diver can and does perform a number of these tasks by feel alone and can work in zero visibility. The vehicles, on the other hand, cannot work without seeing the object on which they are working.

Other aspects of diver work *vis-à-vis* vehicle work must be considered. Rigging tasks with vehicles are generally restricted to attaching a hook or grasping device. The diver cannot only do this, but he can also tie a knot. No manipulator system now in use is known to offer this capability unless the conditions are strictly controlled and favourable. Underwater welding is uniquely the diver's domain. No vehicle today can produce a weld that is in accordance with A.S.M.E. requirements; in fact, there is no reported incidence where vehicle operators offer welding services.

The diver is an incredibly versatile tool, and there seems little prospect of matching his performance with mechanical manipulators. Many of the jobs he does, particularly the scheduled tasks, might be performed through remote mechanical means by redesigning the structures so that they are amenable to mechanical manipulation, but the unscheduled tasks, where something breaks or loosens, will place demands far beyond the present capability of manned or remote-controlled vehicles if the diver is not available for work at 2 km. At this point the likelihood of ambient diving to 2 km should be considered.

The deepest working dive to date was performed by Comex at 309 m in 1975. At the time of writing (March 1977) Comex is scheduled to conduct a 460 m working dive in the Mediterranean. These are record working depths; the average is currently between 90 and 120 m. Undoubtedly 'routine' working dives will be deeper than 120 m, but how much deeper and what is the depth limit are extremely difficult to predict. Some indication of the foreseeable working depth can be gleaned from the diving companies themselves. A sampling of 11 major offshore companies shows a maximum operating depth of 460 m (Ocean Systems Inc., Samson Divers, Comex), the average being 325 m.

Another indication is obtainable from the U.S. Navy, specifically R. C. Bornmann of the Naval Medical Research Institute who projected that compression rate and breathing mixtures for saturation and saturation–excursion divers to 760 m could be available by 1990. Bornmann's estimate time is based on pursuing this goal in an orderly and reasonable manner. Similar estimates from the industrial diving community have not been made public. Various experiments indicate that depths in excess of 760 m are a possibility, but 2 km seems very remote, and personal communications with members of the industrial diving community reveal serious doubts concerning the likelihood of divers working at depths of 2 km. Compression/decompression tables and gas mixtures present just one obstacle on the way to 2 km; others include developing breathing apparatus and environmental protective equipment, determining allowable major contaminant concentrations and developing an ability to treat decompression sickness or any injury or illness that may reasonably occur at 2 km. These are but a few of the foreseeable major obstacles, other may reveal their presence as the depth increases.

It is futile to state categorically that the diver will not proceed safely beyond a particular depth. With the proper resources and no time limitations there is no present way of predicting just what, if any, depth will be an implacable barrier. However, if a time limitation of a decade hence (1987) is assumed, then it seems reasonable that the ambient diver will not have progressed much beyond 760 m. So, from 760 m to 2 km the major problems will arise in support for offshore oil and gas, because the present manned and unmanned vehicles cannot fully substitute for the ambient diver.

DEEP WATER: POTENTIAL PROBLEMS

The magnitude of the potential problems which will confront manned and remotely controlled vehicles at 2 km depth depends upon the tasks they pursue. For example, the problems confronting manned vehicles at greater depths in performing the tasks they now conduct in shallower depths will be small compared with the problems they face as lockout support vehicles at 760 m, and minuscule if they must substitute for the diver at any depth level. The RCVs face similar problems, with the obvious exclusion of diver support, for which they are not designed. For convenience, the following discussion of these problems is categorized thus: general (i.e. problems uniquely imparted by increased depth), diver support, and diver substitute. The problems within each category are not mutually exclusive; some overlapping is unavoidable.

One problem area not discussed below concerns work in the polar regions or under an ice cover. Submersible excursions under ice have been limited, the main reason being safety (i.e. retrieval of the vehicle if it is immobilized) and lack of power to reach areas of interest. While this is not a deep water problem as such, it will be a problem of tremendous magnitude in future attempts to retrieve polar oil and gas resources. Specifically, the major problems are power, navigation and reliability. There is no known project today that is addressing submersible under-ice operations in a concerted and adequately funded effort.

Manned submersibles

General

The general problems associated with increased depth are the predictable consequences of increased pressure.

(a) *Pressure hull configuration*. Below about 760 m depth the shift to a spherical, rather than a cylindrical, pressure hull is now required to obtain the most weight-efficient geometry. The consequence is the least efficient configuration for interior arrangements and human factors.

(b) *Viewing*. All industrial vehicles manufactured in the 1970s have a plastic bow dome. The present domes for vehicles of 240 m depth are 914 mm diameter and 51 mm thick; in 1 km depth vehicles they are 762 mm diameter and 102 mm thick. At 2 km the diameter will be less; consequently, the capability for direct visual observations will decrease.

(c) *Batteries*. Vehicles of 730 m operating depth carry their batteries in pressure-resistant pods on roller trays; this allows for quick turn-around time in replacing spent batteries with charged ones. Greater depth requires stronger and therefore weightier pods. The generally exercised option is to put the batteries in a pressure-compensated fluid to save weight. The consequence is that the batteries must be charged in their containers and quick turn-around time is no longer achievable.

(d) *Electrical interference*. The multi-mission concept for submersibles requires extensive use of electronics, specifically sonar. In some operational modes a variety of equipment and components may be operating concurrently (e.g. propulsion motors; forward-scanning sonar; side-scan sonar; sub-bottom sonar; CO_2 scrubbers; cameras, lights; altitude/depth sonar and navigation systems). Few submersibles are provided with shielded conductors to avoid interference between these components.

The above problems are not particularly difficult to deal with, but they must be brought into the design for efficient 2 km operations.

Diver lockout support

(*a*) *Power*. The major problem in support of the ambient diver will be providing adequate power for heating (hot water and breathing gas) and for employing certain tools requiring electrical power. Studies of the Canadian submersible SDL-1 show that, if the entire energy of the battery supply (36 kWh) could be transferred to the microenvironment of the crew clothing suit with total efficiency, there would not be sufficient energy to provide thermal comfort for the crew of six for 6 h at 0 °C. Underwater welding requires about 25 % more power than welding in air and increases with depth. The Royal Navy Diving Manual specifies a 70–75 V, 300–400 A (30 kW) d.c. generator for welding at about 60 m; only the 30 m long 'Auguste Piccard' can supply this quantity of power. The penalty for increased vehicle size and mass to provide the additional power is discussed below. The alternative of supplying power via a surface umbilical is attractive, but drag and potential entanglement are thereby introduced.

(*b*) *Breathing gases*. The amount of breathing gases for present lockout vehicles limits the actual working time to minutes. Increasing the diver's duration by increasing the amount of gas carried results in a greater vehicle mass. Converting to closed-circuit instead of open-circuit systems can result in a respectable gas saving. An option is possible here as with electrical power – a surface umbilical – but the disadvantages have been mentioned.

(*c*) *Mass and size*. The problems in this instance are introduced by virtue of gaining sufficient payload to supply adequate power and breathing gases to support the diver as discussed above. When vehicle mass (dry) is increased, the repercussions are evidenced in launch and retrieval. There is no doubt that a submersible can be built to supply the required power simply by acting as a battery supply platform, but this vehicle will be extraordinarily heavy and launch and retrieval will be restricted to very large support ships. The recently built Taurus is a 1977 attempt to supply an improved diver lockout and dry transfer vehicle; it weighs 26 t, almost twice the mass of currently operating vehicles. The handling system that must be available to launch and retrieve Taurus will be beyond anything presently in use. Indeed, at 26 t, the various classifying societies' minimal dynamic loading requirements may rule out conventional over-the-stern handling techniques.

Another operational repercussion is introduced by increased size: the potential for damaging the structure being worked upon or inspected. Quite simply, the larger the vehicle the more difficult it is to control. In present diver-supporting rôles the submersible stations itself as closely to the work site as possible; precision manoeuvring and control is a primary requirement. Since most present lockout support vehicles are relatively small, adequate control is obtainable. The large lockout vehicles of the future may not offer adequate control and impacting with the structure is a probability, expecially under fluctuating current conditions. While most impacting would probably damage the submersible more than the structure, the problem is one of safety; but where the structure might be a concrete-coated pipeline, both safety and damage to the coating is jeopardized. Further, placing a large vehicle in the desired proximity for working on a structure may be precluded or severely restricted simply by its bulk.

Diver substitute

Re-examining table 4, which tabulates the capabilities divers now provide to offshore oil and gas, it is seen that a great number (two-thirds) of the tasks a diver performs are also performed by manned submersibles. The table, however, is somewhat misleading and the discussion attending it explains why. In essence, the manned submersible, as an underwater constructor, maintainer or repairer, is extremely limited in where it can work and what work it can do. As a substitute for the diver, the vehicle lacks his agility and manoeuvrability, and its manipulators lack his dexterity and sense of feel. Perhaps the greatest problem is that the submersible itself cannot manoeuvre and hold itself into and around structures where the diver routinely works. Contemporary manned submersibles can support and augment the diver's capabilities, but they cannot substitute for him.

Remotely controlled vehicles

The problems facing RCVs at great depths are somewhat different from their manned counterparts. By taking the human aspect and battery power factors out of the underwater equation, safety is not jeopardized and power is not a limitation. However, the cable which now carries the power and transmits control and data brings with it a new set of problems. The surface ship or platform from which power and control is provided also introduces problems unique to the tethered RCV. The following discussion treats three problematical aspects of the RCV: the support craft; the cable, and the vehicle itself.

The support platform

(a) *Station keeping.* In certain applications, such as long transect bottom surveys or pipeline/cable inspections, the support ship is required to maintain a position directly over the RCV while both are underway. In other applications, e.g. site surveys or hardware inspection, the support ship may be required to maintain position within a limited radius over the RCV. The solution is provided by a support ship with a dynamic position-holding capability, such as bow thrusters and/or laterally trainable stern propulsion. Such ships are available, but they are not often attainable and their cost is high. Furthermore, as witnessed by the recent F-14 search/recovery off Scapa Flow, where deteriorating weather forced the 'Constructor' to abandon the search, they are sea-state limited. Without a dynamic positioning system the support ship may be repeatedly – and literally – blown off station. A conventional two- or three-point static mooring system would solve some station-keeping problems but at the expense of time and a potential for entanglement.

(b) *Launch and retrieval.* The small RCVs are not significantly hampered by launch and retrieval problems, but the larger and more specialized vehicles can and have confronted launch and retrieval problems equal to those of manned submersibles. During sea trials with the M.M.S., described earlier, major repairs were required to the device when its handling system failed and it was dropped on the deck. Calculations for the Hydrotech unmanned pipeline repair system show that launch and retrieval in sea-state 3 is possible; this would equate to a wave height of about 1 m, an extremely calm day in the North Sea and a rarity in the Gulf of Alaska.

The two vehicles mentioned above are quite large; most present RCVs are much smaller. The concept of a multi-purpose RCV requires multi-equipment, increased structural framework strength, greater propulsion and buoyancy: the result must be a heavier, larger vehicle. The operational consequences are for more sophisticated and stronger launch-and-retrieval devices and larger support ships. Several advantages of RCVs, such as use from virtually any ship of opportunity, and ease of transportation, will dwindle as the RCV's size and mass increase.

The cable

(*a*) *Drag.* Hydrodynamic drag on the RCV cable can be provided by vehicle/support ship lateral motion or by water currents, and in some cases by both. The obvious consequences of drag is to reduce vehicle speed; this reduction is acceptable in most operations because high speed is not a primary requirement. The effect of drag on 2 km of cable can be serious if both the support ship and vehicle are attempting to operate while underway, on a pipeline or cable inspection for example. In this application the cable drag could be sufficient to produce a catenary that will pull the vehicle away from the object it is trying to inspect.

(*b*) *Entanglement.* When working around and within a structure the potential for cable entanglement is high; two remotely controlled vehicles have been lost by this means. In one instance the cable fouled in its support ship's propeller and was severed: the vehicle was never found. In another instance the cable fouled in a structure and the emergency cable cutting device was activated. The vehicle apparently surfaced, but it had no surface flashing light or radio beacon and was never located. Twice during an operation in the Santa Barbara Channel a remotely controlled vehicle fouled its cable in the structure it was inspecting; in both instances a manned submersible was launched to recover the vehicle.

(*c*) *Electrical interference.* The problems associated with electrical interference in present RCVs are few because the signal-transmission requirements are relatively simple. However, when consideration is given to the multi-purpose RCV the potential interference problems within the cable can be considerable. The U.S. Navy, during development of Ruws (a 6 km RCV), was forced to develop a cable which employed time-division frequency-division multiplexing techniques for signal and power transmission. The total cost of the cable exceeded $1 M, the production run cost $320 000 for 7315 m of cable. The Ruws itself is still undergoing sea trials; although the electrical interference problems seem to be overcome, final judgement is being reserved until field tests are completed.

The vehicle

In their present rôle as inspection and t.v. photo documentation platforms, an increase in depth should not generate problems which contemporary vehicles find limiting. However, if they are deployed to substitute for the diver, they – like their manned counterparts – cannot provide the manipulative work capability required. In many respects remotely controlled vehicles have the same limitations as manned vehicles: lack of manipulative dexterity and a sense of feel, and a requirement to 'see' the object upon which they are working. Significantly, they demonstrate manoeuvrability which can exceed the diver himself.

TECHNOLOGICAL BREAKTHROUGHS

Many of the problems identified with increased depth do not require technological break-throughs *per se*; instead, they might find solution through a change in design philosophy. However, the loss of the diver's manipulative capabilities is one problem for which no practical substitute or alternative seems to be immediately forthcoming.

A potentially viable alternative to the ambient-pressure diver is the one-atmosphere diving suit called JIM, but for many of the tasks the diver performs the present JIM is too cumbersome and lacks the diver's sophisticated manipulation. Therefore the major technological break-through required to meet the needs of offshore oil and gas at 2 km depth is the development of a manipulator with the dexterity and tactile senses equal to the human hand. This surrogate hand must also be capable of memory and the entire system cannot be much larger in dimension than the human body if it is to find across-the-board application. The task is formidable, and if the solution is required within a decade or less, then alternatives must be considered.

The most readily available alternatives are arbitrarily placed into two areas: operational techniques and design philosophy. The former requires combining present manned and remotely controlled vehicles and one-atmosphere diving suits into a transport/support/work system; the latter requires designing undersea structures and hardware for inspection, maintenance and repair by mechanical manipulators.

Operational techniques

The diver lockout submersible has provided a measure of experience in combining and deploying varied capabilities; the problem remaining is to remove the factor of the ambient pressure diver from the submersible and introduce the RCV and/or the one-atmosphere JIM-type suit. In this combination the three components could work in the following manner:

Manned submersible: transportation of capabilities to work site; provide power and tool/instrument storage.

RCV: perform inspection/documentation and simple manipulative tasks around and within structures where the manned submersible cannot effectively or safely manoeuvre.

JIM-type suit: performs complex manipulative work tasks in confined areas.

The most critical obstacle to obtaining this solution is electrical power. The manned submersible could receive its power from a surface umbilical, but this solution is not altogether satisfactory for it introduces the problems already discussed for the RCV umbilical. Furthermore, it is not a viable alternative to under-ice operations. The pressing need therefore is for an independent power source, such as fuel cells or closed-cycle diesel generators. Other problems can be cited that will result from combining these three capabilities, but if an adequate, self contained power source were available the remaining problems could be readily solved.

To the author's knowledge, there is no government or industrial activity at present which is attempting to combine the three capabilities, therefore the potential problems which would be confronted are pure speculation. The entire spectrum of problems can only be identified by actually combining the manned and remotely controlled vehicle and the one-atmosphere suit and then deploying this system in the field.

Some consideration has been given to the potential of remotely controlled, untethered or robot vehicles in underwater work. Prototypes of these vehicles have been developed and used in the United States and Japan; their capabilities to support offshore oil and gas are meagre. Problems in real-time control, signal transmission, manipulation and power are too overwhelming to consider them practical alternatives to the systems already discussed.

Design philosophy

The 'breakthroughs' of this category are more philosophical than technological. In the early days of offshore oil and gas – because the water depths were shallow – the diver was called in to perform any underwater tasks. Consequently, the hardware used underwater was generally the same as that used on the surface. Bolts, nuts, shackles and rigging techniques were used that were designed for manual manipulation. The location of critical components was often in areas on the structure which were only accessible to the human. This approach is still evident in the design of much contemporary hardware, and it imposes severe limitations on the performance of present manipulators. The performance of present and future vehicles and their manipulation systems can be greatly increased if the designer is aware of their capabilities and limitations, and designs his structure within these constraints. Exxon's M.M.S. is a step in this direction; further assistance towards improved performance of non-diver inspection, maintenance or repair tasks can be provided by considering the factors listed below into the overall design philosophy.

Design components for mechanical manipulation.

Locate critical components in areas accessible to a manned vehicle or an RCV.

Include tracks or rails for guidance and stabilization for the manned or remotely controlled vehicle.

Structures should be as 'clean' as possible to reduce the potential for entanglement.

Legs, braces and other strength members could be visually coded to assist in identification and navigation. Magnetic compasses are all but useless in the proximity of a steel structure.

Quite frequently the fact that a structure is 'unclean' or inaccessible to other than a diver is revealed after installation. Revelations of this nature could prove traumatic if the structure is in 2 km of water. In such instances the design review policy, rather than lack of technology, may be the limiting factor. If, before the design is frozen, it is subjected to an operational analysis review by an operator of a manned or remotely controlled vehicle, it is probable that undesirable aspects of the design would come to light. Such analyses should not only concentrate on installation, but post-installation inspection and testing (visual and non-destructive), maintenance and repair as well. Such a review policy might also serve to identify inadequacies in the undersea vehicles before – not after – they are called upon to assist and would provide an adequate time period to make modifications.

To cope successfully with scheduled and non-scheduled work tasks at 2 km depth requires a cooperative approach. Technological breakthroughs in manipulation and power will greatly increase the performance of manned and remotely controlled vehicles, but these breakthroughs alone will not provide the ultimate answer unless they are accompanied by a closer working relationship between the designer of undersea structures and those who will service them.

Discussion

C. Kuo (*Department of Shipbuilding and Naval Architecture, University of Strathclyde, Glasgow G1 1XH*). I should like to take this opportunity to raise two points.

First, with regard to the rate in which we can close the gap between human and mechanical devices, I agree that this is affected by the rate with which we can develop the technology for manipulation by mechanical devices, but I believe there are other points which must be taken into consideration because they are just as important:

(*a*) We must devise better methods of handling submersibles and large masses through the air–sea interface. At present we are so restricted in mass and size of submersibles that they can only operate up to sea-state 5. However, if a major breakthrough can occur, then the opportunity for taking additional equipment under the sea would open up new opportunities to different approaches to the design and application of mechanical devices.

(*b*) At present the machine tools used are not completely standardized and if we can apply some form of standardization this would reduce the need for the human hand to perform some of the tasks: instead the mechanical devices can take on a more demanding load.

(*c*) The economic implication of trying to copy the human being would be so costly that alternative treatments must be devised.

Secondly, our research tends to throw grave doubts on the wisdom of developing equipment which can produce manipulative dexterity of the mechanical device equal to that of a diver. Our main reasons are as follows:

(i) The more degrees of freedom a device has, the more likely it is to go wrong, and for this reason we should avoid being too sophisticated and limit the methods to fewer degrees of freedom;

(ii) by adopting such forms of standardization it may be possible to relieve a lot of the tasks which need the human hand;

(iii) human divers have limitations and manipulative skill is not necessarily their strongest asset.

I know that Mr Busby is an outstanding diver but I should like to hear his views on these two points.

R. F. BUSBY. Professor Kuo has raised points with which, on the whole, I agree. There are, however, a few thoughts of his upon which I should comment.

Better methods for handling submersibles through the air–sea interface are definitely required. At present the North Sea operators, Vickers Oceanics specifically maintain that they can and have operated in state 7 seas, not state 5 as Professor Kuo reports. While Vickers's performance is a significant achievement, it does not imply that the optimum system for launch and retrieval has been found. It is my belief that the all-weather launch and retrieval system is a large, long-duration support submarine which can deploy a smaller submersible, diver or remotely controlled vehicle while submerged. I think that state 7 is about the limit that one can practically and economically carry out over-the-side launch and retrieval.

In the final analysis, at state 7 and higher the old adage of 'one hand for the owner and one hand for yourself' is a fact of life, and launching and retrieving an object weighing 12 or 15 t is both difficult and perilous. Further improvements will, in my opinion, be best achieved by going beneath the surface rather than staying upon it.

The undersea servicing firms respond to a variety of customers. If they desired to standardize tooling, to which of the many customers would they respond? In virtually every instance the vehicle operator must tool-up to accomodate a particular task and/or a particular piece of hardware. In many instances the task or tool may never be required again. If, as I have suggested, the designers of undersea structures designed components that could be worked on by mechanical manipulation, then it is possible that standard tools could evolve. Until a specific task or tool sees repeated performance or use, standardization is not practical.

Professor Kuo is quite correct in remarking that the economic implication of trying to copy the human hand would be costly. I have not recommended that this be a design goal. My intent was simply to point out that the lack of this capability will severely limit deep-water support of oil and gas in the forthcoming decade. Co-operation on the part of the undersea structures designer, as I have mentioned previously, could reduce the technical capabilities requirements of a manipulator system. As an example, one can, with perseverance, proper tools and adequate lighting, open a knobless door, but a latch or knob certainly reduces the time and effort.

M. W. THRING (*Queen Mary College, University of London, Mile End Road, London E*1 4*NS*). Mr Busby referred to the need for mechanical hands which can have sufficient sensory feedback to tie a knot and also to have the memory of a human being. Such hands can be provided by the technology of telechirics (hands at a distance). These have already been developed for work inside nuclear reactors with all the movements of the human arm and at least one grasping movement, together with sufficient force of sensory feedback.

I am already working on the possibility of using telechirics so that a miner can use all his skill down a mine while remaining on the surface. In this case he will also need feedback of binocular vision, so that he can use his hands exactly as he would if he could see the work he was doing, and feeling it with all his trained skill.

It would certainly be possible to develop telechiric hands for work at and depth under the sea within 5 years if we really put our minds to it. These hands could be operated either by a man inside a submersible or attached to a suitable vehicle operated by cable from a ship. In this case the cable could also carry the power supply. Visual feedback can be by binocular t.v. cameras and a very powerful light source movable very close to where the hands are working, or by sonar, but in any case the tactile and force feedback would so so good that the man could do the job by touch if necessary.

R. F. BUSBY. The concept of transferring nuclear manipulative or telechiric techniques to undersea work is not new. In the early 1960s few discussions of manipulative capabilities were completed without questioning why the nuclear techniques were not employed on submersibles. I cannot argue Professor Thring's thesis that telechiric hands for work at any depth could be developed within five years if we really put our minds to it. But there are other technical and deployment considerations which are not readily apparent. Each finger of the human hand has not only sensory perception but memory as well. Technological memory is the domain of the computer. Further, the human is deployed by a 'platform' with outstanding characteristics for manoeuvrability and obtaining stability in close spaces. While I do agree that the human hand can be duplicated in the laboratory to a remarkable degree, I do not believe that five years is enough time to develop the techniques for its deployment within the space, manoeuvring and power constraints of contemporary manned and remote controlled vehicles.

To duplicate the human hand the telechiric system must at times operate in waters of zero visibility. Granted that acoustic imaging has made great strides in recent years, but as a replacement for the human hand's sense and memory it is a long way off the mark.

In view of these technological and operating constraints I cannot see complete duplication of the diver's manipulative dexterity by 1982, and still see this as the single major technological obstacle to providing deep water oil and gas support.

Phil. Trans. R. Soc. Lond. A. **290**, 153–159 (1978) [153]

Printed in Great Britain

The underwater contractor: his rôle and development

By R. E. England

Vickers Limited, Offshore Engineering Group, Banda House, Cambridge Grove,
Hammersmith, London W6 0LN, U.K.

Recovery of hydrocarbons from the seabed has a long history, particularly in the Gulf of Mexico and Middle East.

Steps and processes required to extract oil from the seabed are essentially the same as those for recovery on land, but with the additional burden of bridging water depth.

The transition to exploitation of North Sea discoveries brought problems not always fully appreciated, the most persistent of which is still the unpredictable and harsh weather and sea conditions. The rate of exploitation and exploration depended upon the work speed in the more favourable surface conditions.

Experience and ingenuity have contributed greatly to reducing downtime. However, worldwide moves into greater depths and even less sheltered waters, stretch the suitability and cost-effectiveness of conventional structures and the techniques employed for seabed equipment maintenance and inspection.

Consequently, the recently developed concepts for installation and servicing of engineering complexes on the seabed are providing one of the most significant steps towards the goal of year-round operations.

1. The growth of the underwater contractor

In his opening remarks Dr Birks (1978, this volume) underlined the fact that the oil industry in the last five years has nearly doubled the depth of water in which it is exploring. It has much further to go yet.

Offshore hydrocarbon recovery over the last 25 years has introduced a new and highly specialized sphere of contracting. Compared with the way in which land-based operations developed, its growth and complexity have been both considerable and rapid. Pressures upon this development have increased also because of the political and economic price of oil in the North Sea at this point in British history.

The first steel structures offshore appeared in 7 m of water in the Gulf of Mexico in 1947, but by 1970 large structures in 150 m of water were becoming the norm. While offshore oil accounts for some 19 % of the total world's crude oil supplies with North Sea yielding approximately 4 %, the investment necessary must be regarded as enormous by any standards. By 1985 it is expected that offshore rigs will contribute 35 %. This gives some indication of the problems which must be faced and the investments which will be necessary. Work began in the North Sea in the Southern sector gas fields, followed by the oil fields Montrose, Forties, Brent and Ekofisk. The North Sea thus became well established as a development and production area. To meet these requirements a battery of specialized tools was necessary.

The almost overwhelmingly difficult environment, deep and exposed water, with high wind and wave conditions, were responsible for considerable delays and, indeed, will continue to tax the engineer's capability and ingenuity for years to come. Nowhere had such conditions previously been experienced, but the quality of the oil was considered to be excellent and its proximity to the market combined to offer great potential. In the future, oil production will move further north into still deeper and rougher conditions (including ice).

Oil-company financing of these vast undertakings requires the earliest possible returns from oil revenue and great pressure is brought to bear on contractors to obtain the earliest possible production and cash flow. In their turn these pressures provide the subsea contractor with commercial opportunities over this whole new field. The offshore and the underwater contractor in particular, are therefore presented with a challenging if daunting range of opportunities for investment and opportunities in the future. British industry really must take advantage of the North Sea to develop capabilities for later worldwide exploitation.

2. What the underwater contractor does

It will be appreciated that oil companies' primary interest is the end product and not the means of obtaining it. The cost of offshore hydrocarbon production is now so great that the resources of the oil companies are stretched to the limit to provide the necessary finance. Therefore, while the client oil company must identify his problems and his development direction, in his own interests it should encourage the contracting industry to develop and provide the means. Most contractors are hard-headed, down-to-earth 'doers' and forward thinking and development is not a way of life for them. Deep-water offshore work is much akin to space technology: virtually everything has to be undertaken in a hostile environment. Good science, good engineering and above all good safety and management of difficult operations are essential. The underwater contractor will rely heavily on vehicle and systems to be able to do his job safely and efficiently. The client oil company therefore frequently engages a project management contractor whose task is to manage the whole project on behalf of the client engaging, monitoring and managing specialized subcontractors to achieve the overall objective.

A total field management project would commence with feasibility studies covering all aspects of the development, including projected programmes, time scales, cost breakdowns, etc. Thereafter the contractor initiates, coordinates and oversees all design, manufacturing and installation services of the selected subcontract entities on behalf of the oil company.

The management of a programme from early subsystem development through to operation of a deep water field production demands a unique approach, balancing the risks incurred by new developments against the advantages of introducing them.

Some of these problems and their evaluation are of such magnitude that oil company/ industry/Government/E.E.C. funding is essential.

The management of North Sea engineering requires a considerable degree of innovation which must be continually reassessed and fed back into an ongoing programme for future field equipment. It is equally important to ensure that, where subsystem development is taking place in discrete packages, each is pursuing the same long-term objective and this point is taken up later. The offshore and underwater engineering industries can learn much from aerospace and warship engineering traditions. However, the commercial end objective is usually seen in stark clarity and therefore in offshore operations empiricism and pragmatic solutions have an important rôle to play.

Management must constantly assess forecasts of resources, market requirement and development scenarios, the postulation of programmes to produce future engineering systems, and the identification of the key system elements or building bricks of the system; all this is aimed at making production under extreme conditions economically feasible and more cost-effective and commercially attractive.

Inevitably every application and oil field is different and a solution depends upon the precise specification for the particular field. In considering the strategy to be adopted therefore, the following criteria are dominant:

(a) Technical and financial risk to be minimized.

(b) The total package resulting from any innovation to be cost-effective to the industry.

(c) Government legislation and requirements to be met.

(d) Requirements of certification bodies to be satisfied.

Technological feasibility is a major challenge put increasingly to the test with the move into deeper waters. As a result, both technology and management are being stretched to their limits.

The oil company and its project management contractor have to work in the closest possible harmony and the latter is to be considered as an extension of the oil company's organization. There are, however, many specializations within the whole which call for expertise that can only be provided by organizations, individuals and teams offering tailored services. Both the oil company and the contractor, therefore, face immense investment risks but I would venture to remind oil companies that they will enjoy the profits of the end product – oil – in what can, in the longer term, only be seen as an expanding energy market.

A small sample of some of the areas in which the subsea contractor becomes involved is as follows:

   (i) general area survey;

  (ii) detailed block/area survey work, especially soil mechanics;

 (iii) pipe route surveying;

 (iv) provision of subsea facilities including wellhead, possibly encapsulation thereof, manifolds, riser, export line, flowline, satellite well, anchoring, pipeline trenching/de-trenching, subsea template, platform positioning;

  (v) precise seabed positioning and measurement;

 (vi) hook up;

 (vii) scour/sand build up surveys;

(viii) manned wet/dry intervention system;

 (ix) diver lockout maintenance;

  (x) debris clearance;

 (xi) inspection, maintenance and upkeep;

 (xii) cathodic protection check;

(xiii) leak detection;

(xiv) photography;

 (xv) equipment recovery at field close down.

The complexities, development and costs that flow from these activities are highlighted in one operating unit of my own Offshore Group. Analysis showed that to reach an objective in only four of the above subsea market areas, 96 development programmes would have to be initiated with costs ranging between £2000 and £2 000 000 (and this does not include the basic delivery vehicle).

3. A TYPICAL NORTH SEA DEEP-WATER FIELD DEVELOPMENT
PROGRAMME AND ITS SUBSEA ELEMENTS

A typical North Sea deep-water production programme can take some 6 years to mature. However, as already mentioned, the oil company and the offshore subsea contractor seek ways of reducing this time, of cutting down costs and reducing the effects of weather. But the deep-sea underwater environment presents a number of difficulties: pressure, cold, visibility, navigation, provision of power, and surface weather; all of which increase costs and stretch technology.

The floating platform (in which mass will be a limiting factor) is one way in which cost-effectiveness is being improved. More and more equipment will be placed on the seabed to reduce weather downtime. The problem of the surface/seabed pipe connection or riser is currently receiving a good deal of attention.

Seabed drilling is terminated in a wellhead or 'Christmas tree' which is substantially similar to its land-based counterpart except for additional complications, such as temporary and permanent guide bases, remote operated control, flowline loops and connections. Wellhead equipment has to be wireguided from the surface and connected either remotely or using divers or submersibles, and then there is the problem of maintaining them.

The American Lockheed Company offer one solution with their subsea wellhead cellar, serviced from a diving bell or transit capsule from a platform or rig. A British solution for wellheads not necessarily under the platform is the submersible accessed subsea chamber from Vickers-Intertek. The system provides a one-atmosphere pressure chamber, but unlike the Lockheed cellar it remains a wet environment. Water is retained in order to simplify the structure required to reduce the risk of gas contamination, fire or explosion on the seabed, and to reduce power requirements. Access is gained via a proven technique of transferring men by transfer submersible to the seabed chamber The submersible lands on top of the transfer hatch of the chamber and secures itself without mechanical links by reducing the water pressure in the submersible's underwater transfer 'skirt'. Only a minimal amount of power is required to remove the small quantity of water needed to reduce the pressure in the seabed chamber from ambient to one atmosphere. The intervening hatch can then be opened without fear of the submersible's flooding, and maintenance operators who do not need to be trained divers can enter the system. The submersible provides power for lighting and a workshop facility for the divers to work in a one-atmosphere pressure environment.

A great advantage of both of the one-atmosphere systems is avoidance of risky, time-consuming and costly saturation diving methods necessitating long periods of diver decompression, although there is an ultimate depth restriction.

Offshore oil production involves the use of 'satellite' or 'step out' wells in addition to those underneath a platform. To avoid proliferation of platforms, it is necessary to connect these wells into a manifold and thence to the platform riser. Considerable pipework and cable interconnection on the seabed is necessary, together with the means of laying, hauling in and making final connections. These are currently difficult problems and subsea encapsulations, habitats, chambers and the submersible have a part to play. Pipes and cables must be buried to protect them from damage by deep trawling and ships' anchors.

The subsea manifold centre acts as a collecting point from the satellite wells, for production testing, entering through-flowline (t.f.l.) tools, and export of oil to tanker loading buoys or ashore. The system is designed to be unattended in the main, with manned intervention only for installation and maintenance.

An Exxon subsea production system, currently operating as a test installation in the Gulf of Mexico, makes use of an electrohydraulic supervisory control system. Pumpdown t.f.l. tools are used for servicing, and a manipulator system, operated remotely from the surface, undertakes maintenance of the equipment which is installed in modular configuration.

A subsea completion under a floating platform would comprise the following (this concept is one of those engineered by my own company for a major oil company):

   (i) a disposable drilling template;

   (ii) encapsulated wellheads;

   (iii) flowline connection from local and satellite wells into manifold chambers;

   (iv) manifold chamber containing pipes, valves and controls in inert gas, having an accessible control chamber at one end and looking like a piece of nuclear submarine;

   (v) connections to surface by riser/umbilicals.

At one point or another, all the oil and gas extracted from the seabed has to pass through a pipeline on its way to the shore. At any moment, from the time it is being laid onwards, a pipeline may need to be repaired. There is no alternative but to replace the damaged section, at present by mechanical means or by welding, with the use of divers. Both techniques have been used successfully in the North Sea. The contractor has to remove the surrounding seabed material to expose the pipe. The protective coating has then to be removed, the damaged section of pipe cut away and the pipe ends treated in readiness for re-jointing. The new section of pipe then has to be transported to the site and manoeuvred and held precisely in position, while the repair is effected.

As always, water depth plays an important part in the repairs. Conventional underwater wet-welding techniques become very unreliable as depth increases beyond 30–40 m due to the quenching action of the water. One solution has been to install a hyperbaric underwater welding habitat about the pipeline. A mechanical repair technique under development is that of Hydro-Tech in which hydraulically operated ball joints are inserted over the pipe ends.

Explosive welding is another method being developed. To date, consistently successful underwater welds are being achieved with 8-in and 32-in diameter pipes at depths to 120 m. The aim of this system is to dispense with divers entirely and to operate the system remotely from an attendant submersible. Welding actually occurs when an explosive charge inserted within the pipe forces together the individual plates with very high velocity. Structural analysis shows the weld area to be subsequently stronger than the parent metal.

Each of the jointing applications described can be and is being used also for the underwater connection of tie-ins, etc.

The submersible has played a vital rôle in the offshore scene since the 1960s. Much effort has been devoted to the development of the military submarine and we have come a long way from Foulton's submarine 'Turtle' of 1770! The world's first practical underwater submersible system harnessed to offshore engineering was pioneered by Vickers Oceanics in 1969 and now forms the basis of many deepsea operations. The versatility and dexterity of man as a diver will be difficult to replace. Sophisticated high-pressure-resistant diving suits like JIM stretch the diver's capability further. By the same token, the scope of work now being undertaken on the seabed could never have been achieved without the submersible.

Acceptance of the submersible for underwater work was a very long and slow process, but it is now seen to be the practical solution to a number of deep and shallow water problems. Typical submersible applications include:

one-atmosphere personnel transfer underwater;

pipeline route and platform site surveys;

pipeline inspection, burying/deburying and repair;

diver lockout services;

platform structural surveys and non-destructive testing.

With the exception of divers 'locked out' of submersibles, no decompression of the crew is required because of the retention of one-atmosphere pressure within the crew compartments. Power for the motors and manipulators is provided by batteries, which at present is one of the limitations of untethered manned submersibles. Efforts to improve battery performance or a successful alternative power source are a constant theme of research and development.

In the present submersible the battery mass is typically about 1.5 t representing some 10–15% of displacement – this provides some 50 kW h of energy capacity. Underwater power for tasks in the order of 15–20 kW are already encountered implying capacities of 120 kW h for 6–7 h dives. Other power sources investigated include the H.T.P. turbine (but we already have enough problems!), the I.P.N. turbine (pressure degrades performance, it pollutes and it is noisy), the re-cycle diesel engine, the Stirling engine, and the Alsthom fuel cell; all of which require further development. The need is great enough to suggest that substantially higher power sources for the submersibles or submarines of the future can be expected.

The recent introduction of a high-strength, high-technology glass-fibre reinforced plastic hull by Vickers has brought about savings in mass and maintenance tasks and given the benefits of additional buoyancy and power reserves.

The application of submersibles for diver lockout work enables the diver to be taken directly to his work site and once outside the submersible his life support, power for tooling, light source and supervision from within the clear-thinking one-atmosphere environment, are all provided by the submersible. This system is less prone than diving chambers to adverse surface conditions.

A special closed-circuit heating system to combat cold recently enabled a diver to establish one of the longest dives in the North Sea. He worked outside a submersible for $3\frac{1}{2}$ h at a depth of 467 ft at the Piper 'A' platform, in a joint Vickers/Oceaneering International effort.

As an alternative to manned submersibles a number of companies are investigating the potential of unmanned submersibles. Advantages include the safety aspect of dispensing with the need for men underwater, extra power for longer periods due to power supply via umbilical cable and a better facility than manned submersibles for operating from ships of opportunity, as opposed to purpose-built ships.

Although unmanned submersibles have proved their ability to undertake underwater tasks, they only contribute to the total answer. They may prove most useful in the rôle of underwater inspection of large fixed structures. Typical problems include a high degree of drag on the umbilical cable and hence a reduction in performance capability, lack of definition by the television guidance system and, like the manned submersible, in the shallower end of the market where divers can operate, lack of manual dexterity.

Summary

This paper has outlined some of the main areas of challenge facing the contractor at this time, crudely divided into two parts:

(1) the total overall project engineering and management activity;

(2) provision of specialist services and products.

While the first rests very largely at the door of the oil company and his appointed contractor, the second presents a bewildering selection of opportunities and pitfalls for the subsea contractor. Each avenue he explores opens up tens of further alternatives, presenting development-decision trees of substantial, and sometimes frightening, complexity. The engineering, marine and construction industries are already part of the way along the road leading to technical advances akin to those made in man's conquest of space. The speed with which solutions must be found is, however, a matter of the commercial targets which must be attained rather than national prestige. Management in these industries will have to grasp offshore opportunities for the U.K. Client oil companies and the government will need to play their part too in encouraging developmental investment.

In the pursuit of development, millions of pounds can be poured into large specific projects, when often the solution lies in smaller but related steps within a larger overall plan – in other words the identification of 'building blocks' leading continuously onwards and upwards in technology. Vickers development of the submersible *system* clearly demonstrates the method. The *vehicle* existed but the systems to deploy it, control it, give it suitable tools and recover it, were missing. Offshore subsea engineering will be a constant repeat of this situation.

Multi-sponsored projects involving E.E.C./Government/oil companies/contractors/suppliers are an attractive answer to funding innovative development: they can also lead on to partnerships and joint agreements which provide the resources necessary to supply the complementary services which will be required by North Sea and later worldwide offshore hydrocarbon and mining production.

The subsea contractor has made a remarkable contribution to sea-floor development – and from here on the pace will quicken. He will need to bring all the sciences to bear if he is to play his part in moving into deeper water. Not the least of these sciences will be the management science. He also must have a practical, common sense and healthy respect for the sea.

Phil. Trans. R. Soc. Lond. A. **290**, 161–177 (1978)   [161]
Printed in Great Britain

Effect of the environment on processing and handling materials at sea

By T. H. Hughes

*Mineral Processing Division, Warren Spring Laboratory,
Stevenage, Herts, SG1 2BX, U.K.*

[Plates 1–4]

Compared with conventional land-based mining and processing operations, the exploitation of minerals from the seabed, particularly in deep water, involves a vast range of new problems in conducting the various stages of mining, transportation, processing and disposal of waste products, in a marine environment.

In all such operations the ways in which local sea and weather conditions and their seasonal variations affect the stability of the vehicle, be it ship or other floating structure or submersible from which the operations are being conducted, have to be taken into account. The resulting motion together with vibration generated by propulsion and other machinery are significant factors in the performance and behaviour of equipment and materials during processing, handling and transportation operations at sea. In deep-sea mining operations at depths of 2–5 km the effects of associated pressure, salinity and temperature must also be dealt with.

The paper reviews the present state of such knowledge as currently practised in continental-shelf operations, and as proposed in various deep-sea mining operations. Associated research requirements for future mineral exploitation in the deep-sea environment are discussed.

Introduction

During the past decade there has been world-wide interest in the sea bed and the ocean deeps as a source of raw materials to meet ever increasing demands or to replace exhausted land-based resources. Earlier papers in this volume have illustrated the tremendous advances and new developments in marine technology which have led to the successful exploitation of offshore oil and gas. Although there has also been widespread prospecting and associated technological activities towards the exploitation of marine minerals, the rate of development by the mineral industries has in general been considerably less sensational than that of the offshore oil and gas industries. This is due to constraints imposed as much by economic factors as by technological problems. For example, the overall viability of an offshore mineral operation must be related to the cost of exploiting equivalent land-based mineral resources, world market values and demand. In deep-sea mining operations, there also remains a lack of agreement on International Law of the Sea relating to the security and rights of concessions.

A number of minerals, including sand and gravel, tin, iron, titaniferous sands (in near-shore operations), calcium carbonate in various forms, barytes and sulphur, are currently being recovered from the sea bed in commercial-scale operations. With the exception of barytes and sulphur, all these operations are carried out by dredging. Inevitably, as near-shore deposits become worked out, future trends and technology must be established for operations to move further and further out to sea, dredging at greater depth and consequently in worsening sea conditions. In this situation there is considerable incentive towards primary processing operations at sea to avoid the need for transportation of large quantities of waste materials which can be rejected on site.

It is recognized that compared with conventional land-based mining and processing operations, the exploitation of minerals from the sea bed, particularly in deep water, involves a
wide variety of new problems, including those arising from local sea and weather conditions,
their seasonal variations and their effect on the vehicle, be it a ship or other floating structure
or submersible, from which the operations are being conducted.

The resulting motion, and vibration generated from propulsion and other machinery, are
significant factors, not only in the mining and lifting of material to the surface, but also on
the subsequent behaviour of equipment and material during processing, handling and transportation operations at sea. In deep-sea mining operations at depths of 2–5 km the effects of
associated pressure, salinity and temperature must also be dealt with.

This paper first reviews the present state of such knowledge as currently practised in mineral
operations on the continental shelf and considers what future development trends are likely
as these operations move into deeper water. In the continental-shelf operations, marine sand
and gravel and marine mineral mining (typically and principally represented by alluvial tin)
present distinct problems and are treated separately. Following this, corresponding aspects of
proposed operations in the exploitation of deep-sea minerals are dealt with. Finally, associated
current and future research requirements relating to all these activities are discussed.

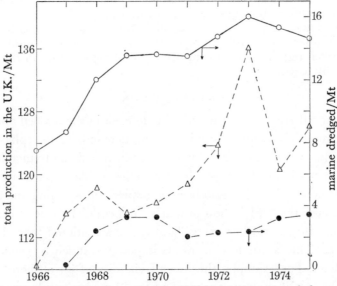

FIGURE 1. Trends in production of sand and gravel in the U.K. over a 10-year period: △, total production in the
U.K.; ○, total marine dredged; ●, export marine dredged.

MARINE SAND AND GRAVEL

The U.K. marine sand and gravel industry is one of the largest offshore mining operations
of its kind in the world. It produces various grades of sand and aggregate which are used mainly
by the building and construction industries. A maximum annual production of 16 Mt was
achieved in 1973 (Institute of Geological Sciences 1977). This corresponds to about 13 % of
that produced from land-based sources and is estimated to be a similar percentage of the total
marine aggregates produced throughout the world. Since then, the decline in the requirements
of the building and construction industries in the U.K. has resulted in a reduced market

demand and a shift in the proportion exported to other European markets. The growth rate and trends in these requirements are illustrated in figure 1.

(a) Dredging methods

The industry recovers the sand and gravel from the seabed by dredging vessels which vary in size and cargo capacity up to about 7000 tonnes. A typical sand and gravel dredger is shown in figure 2, plate 1.

Suction dredging is almost universally used. In this, the aggregate is transported up a dredge pipe from the sea bed to the vessel from depths of down to about 50 m. The flow is induced either from a centrifugal pump located as low in the ship as possible or from a jet lift system in which a negative pressure is created by high-pressure jets located near the lower end of the dredge pipe. Jet-assisted centrifugal pump systems are also used.

Depending on the type of deposit being worked, either hole digging or tail dredging can be adopted. Hole digging uses a forward-pointing dredge pipe, usually for working deep deposits with the ship at anchor. Trail dredging, as the name suggests, uses an aft-facing dredge pipe. It is usually used for working shallow deposits, whilst the ship is just making way, but can also be used for deep deposits with the ship at anchor.

(b) Shipboard motion and compensating systems

Movement of the ship in relation to the seabed is one of the key factors relating to the limits of sea conditions under which it is possible to operate the dredge. A swell compensator is used to remove some of the relative motion between the ship and the head of the dredge pipe. It is a passive type of hydro-pneumatic system which takes in or pays out the wire rope supporting the dredge pipe according to the load on the rope. The pressure in the system is adjustable to give the required level of contact pressure between the dredge head and the seabed. Thus as the ship starts to lift the dredge-head off the bottom, the load on the rope increases and the compensator pays out more rope until contact pressure is restored. Conversely, when the ship sinks into a trough, the load decreases and the rope is pulled in. However, because the compensator has only a limited operating range it is not always able to provide complete compensation for the motion of the ship in heavy seas. In this situation, a very high sense of anticipation is required by the dredgemaster in manually assisted control of the winches, to enable dredging to be carried out without risk of damage. Typical motion data recorded on a dredger operating in conditions of up to Force 5–6 on the Beaufort scale are shown in figure 3.

(c) On-board processing

The main processing operation which takes place on board the dredger is screening, to reject either the sand as undersize or the gravel as oversize, depending on whether gravel or sand is the required marketable product. In either case, a separation at $\pm \frac{3}{16}$ in (5 mm) is usually aimed for. The screening plant may range from a single static inclined-screen deck (figure 4, plate 1), to either two or four vibrating-screen decks. Where two or more screen decks are employed a feed distributor is needed to ensure uniform volume flow, solids concentration and particle size distribution to each screen deck. Investigations at Warren Spring Laboratory (W.S.L.) have shown that the design of the distribution system is of considerable importance in achieving these aims since the uniformity of volume flow and solids concentration distribution is significantly affected by the configuration of the pipework leading to the distributor. However, the effect of shipboard motion does not appear to alter its overall performance.

(d) Future trends

One of the recommendations of the Report of the Advisory Committee on Aggregates, *The way ahead* (Department of the Environment 1975, §6.35), was that the Department of Industry, in consultation with the Department of the Environment, should encourage research into improved dredging technology, and in particular dredgers capable of dredging in deeper waters (in excess of 36 m) and methods of moving seabed overburden which overlies gravel.

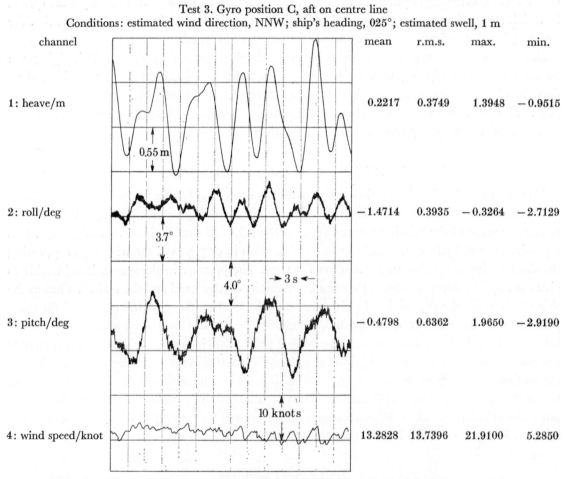

Test 3. Gyro position C, aft on centre line
Conditions: estimated wind direction, NNW; ship's heading, 025°; estimated swell, 1 m

channel	mean	r.m.s.	max.	min.
1: heave/m	0.2217	0.3749	1.3948	−0.9515
2: roll/deg	−1.4714	0.3935	−0.3264	−2.7129
3: pitch/deg	−0.4798	0.6362	1.9650	−2.9190
4: wind speed/knot	13.2828	13.7396	21.9100	5.2850

FIGURE 3. Typical motion data recorded on sand and gravel dredger.

In the present climate of economic constraints, there is little incentive for the industry by itself to conduct development work towards immediate achievement of these aims. Nevertheless, a limited amount of work has already been carried out in this area in terms of design and feasibility studies in which various systems have been considered. These have included suction systems with submerged pumps (G.E.C. Mechanical Handling Ltd 1973) on rigid and flexible dredge pipes, air lift systems, vertical suction pipes with heave compensation similar to that used on drill ships, submersible dredge pods and barges with buoyancy tanks for lifting them to the surface when filled, and underwater processing facilities for rejecting unwanted material

FIGURE 2. Sand and gravel dredger operating in rough seas.

FIGURE 4. Screening plant on board a sand and gravel dredger.

FIGURE 5. Sectional model of a tin dredge.

FIGURE 6. Tin dredge in an offshore operation.

Phil. Trans. R. Soc. Lond. A, volume 290

Hughes, plate 2

FIGURE 7. Illustration of cargo liquefaction originating from the centre.

Figure 10. Flotation tests on ship's motion simulator, with froth discharge in line with maximum roll motion. (*a, b*) Simulator in two attitudes of angular motion; (*c*) overspill with cell tilted forwards; (*d*) no discharge with cell tilted backwards.

Phil. Trans. R. Soc. Lond. A, volume 290

Hughes, plate 4

FIGURE 12. Cargo stability testing on largest simulator platform.

on the seabed. Although the technical feasibility of such systems may have been established, further development of the preferred ones largely depends on costs, market demand and resources. The basic intrinsic value of the product (about £1.75/t) dictates that only relatively cheap and simple systems can be considered compared, for example, with those which might be commercially viable in offshore oil and gas activities.

Market demand depends on the overall economic climate in the U.K. and European countries bordering the English Channel and Southern North Sea, and the corresponding growth rate in their building and construction industries. Future availability of resources is related to demand and thus to the rate at which existing near-shore deposits are used up. Another factor which has to be taken into account is the proportion of these deposits which may be 'sterilized' by fishing and other environmental considerations. An alternative possibility in the medium term is the removal of contaminants such as chalk, shell and clay. Available survey information estimates that approximately 200 Mt of economically extractable gravel, contaminated with small quantities of chalk, exists in U.K. waters.

Finally, the problems of removal and disposal of overburden overlying deep gravel deposits and their consequent effects with respect to erosion and the ecology of the marine environment have to be resolved.

OFFSHORE ALLUVIAL TIN

Alluvial tin dredging dates back to the early exploitation of inland placer deposits in Malaya about 1913. Since then, the tin dredging industry has developed widely into operations on deposits in riverbeds, estuaries and lagoons in many areas of SE Asia and, more recently, has extended into offshore areas of Thailand and Indonesia. Cassiterite (SnO_2) was probably the first metalliferous mineral to be exploited from the seabed and, up to the present, tin remains the only non-ferrous metal commercially produced by the marine mining industry. The most recent statistics for tin production, on the basis of tin in concentrates, estimates that world production from all sources in 1975 was of the order of 175 kt, of which about 13 % was dredged from estuarine or inland sources and 7 % from offshore dredging operations (International Tin Council 1975).

In relation to overall tin production, the offshore contribution is likely to become greater in the future in the light of prospecting data presented to the Fourth World Conference on Tin, in November 1974, for the offshore areas of Thailand, Malaysia and Indonesia.

Tin dredging operations may resume in the St Ives Bay area of Cornwall in 1977 (Beckmann 1975). It was originally planned that a mobile 'walking platform' be used as a base for the mining and processing plant but more recent information indicates that a dredging vessel will be used for the initial operations.

(a) Tin dredges

A tin dredge has little in common with either harbour-clearance or sand and gravel dredgers. It consists primarily of a non-propelled pontoon on to which is built the necessary superstructure for supporting the digging arm, the screening plant for removing the coarse stones, etc., from the feed, and the mineral separation plant which also includes a feed distribution system and a tailings disposal system. The dredge is positioned and fed into the digging face by means of a series of winch-operated mooring lines. A modern dredge can handle up to 1150 m^3 h^{-1} and operate at depths of 50 m. The general construction of a tin dredge is illustrated in figures 5 and 6, plate 1.

In offshore operations the effect of swell has a significant effect on the overall operation, not only in the stresses imposed on the digging arm but also on the performance of the treatment plant. A number of developments to overcome these problems have been considered including submerged plates to stabilize the pontoon, increasing the length of the pontoon to over-span the maximum expected wavelength, and semi-submersible structures and swell-compensating systems similar to those used on sand and gravel dredgers.

(i) *Bucket-ladder dredges*

These have been used for the past 60 years with few changes in basic concept except for improvements in the pivoting and suspension systems and in articulation of the toe. The mechanical stresses imposed under conditions of swell are very large due to the fact that the ladder assembly may weigh up to 1200 t, particularly when digging near bed rock where the tin concentration is invariably at its maximum.

(ii) *Suction-cutter dredges*

These consist of a cutter head rotating at the end of a ladder or framework. The rotating axis of the cutter is in line with the ladder and discharges into a submerged centrifugal pump also mounted on the ladder. It has the advantage that the supported weight is considerably less than that of a bucket ladder dredge, but for alluvial mining, where the mineral values are not evenly distributed in the material being dug (i.e. deposited on the bed rock), the existing cutters have difficulty in properly cleaning the bed rock and produce wide variations in the solids concentration in the feed to the treatment plant. The same may apply to bucket wheel dredges, as yet not fully developed for this application.

(iii) *Screens*

Trommel screens are almost universally used on tin dredges where the primary objectives in design are maximum capacity with minimum space requirement.

(iv) *Mineral separation plant*

Good distribution of the screen-undersize slurry to the treatment plant has always been a problem on tin dredges. Numerous systems have been tried in recent years. By far the most satisfactory system of distribution to have emerged in the tin dredging industry is a splitter system of distribution, by which the feed to concentrating units is controlled through self-compensating triangular orifices. An important feature of this system is that as well as giving reasonably even flow rates to all units of the treatment plant, it also ensures an even grade of feed. Experiments at W.S.L. (Hughes & Joy 1972) have also shown that this device maintains good distribution performance even under extreme ship-board motion.

The method of mineral separation universally accepted in the tin dredging industry is by jigging, which is one of the oldest of gravity separation techniques. It works on the principle that vertical pulsations of water through a bed of particles will cause separation and stratification of these particles on the basis of their specific gravities. The heavier particles work to the lower part of the bed while the lighter particles are forced to the top, the products being collected separately. In a modern plant this may take the form of either conventional rectangular jigs, 1.07×1.22 m, or radial jigs 7.62 m in diameter. Various opinions exist in the industry on the capacity and performance merits of these two types of jig with respect to cost and space requirements.

(v) *Tailings disposal*

The waste tailings are disposed of either by pumping or gravity discharge through sand chutes designed to convey them as far as possible behind the dredge and away from the working face of the deposit. With the increasing concern of government departments on the effects of mining generally on the environment, ever more stringent controls are likely to be imposed upon tailings disposal to regulate rates of discharge, size distribution and methods of deposition with respect to tidal flows, prevailing currents, etc.

(b) *Current designs of tin dredges*

At present, two new tin dredges are under construction, to commence operation in offshore areas of Indonesia on completion. One of these is reportedly the largest offshore tin dredge ever built (Anon. 1977): $110 \times 30 \times 6.5$ m with a bucket capacity of 0.85 m³ for dredging at a nominal rate of 1836 m³ h⁻¹ down to a depth of 45 m.

To achieve a high degree of availability in the prevailing wind and sea conditions on site, it is provided with a hydro-pneumatic compensating system in the bucket-ladder suspension. The mineral concentration plant consists of two revolving screens, circular primary jig and rectangular secondary, tertiary and clean-up jigs.

However, other schools of thought in the operating industry favour simplicity in design and maintenance with maximum reliability. The second dredge has been designed to meet these specific requirements. It is $108 \times 32 \times 4.75$ m with a smaller bucket capacity of 0.625 m³ for dredging at a lower nominal rate of 975 m³ h⁻¹ but at an increased depth of 50 m (Hewitt 1977). A special feature in its design is provision to obtain optimum fill of the buckets over a wide range of angle at the digging face. The treatment plant consists of two revolving screens and again, illustrating differences of opinion, rectangular type jigs are used throughout for the concentrating operations.

(c) *Future trends*

Although current designs of bucket-ladder dredges are generally limited to an operating depth of 50 m, design studies have been carried out for depths of down to 60 m. Limiting factors are bearing design for the main pivot and for the suspension system in view of the enormous basic load imposed by the extended ladder assembly plus shock load safety factors under open sea conditions.

For the future, it seems likely that some departure from conventional designs will be necessary for operations at depths greater than 60 m. For example, it is believed that deep tin deposits exist in the straits of Malacca which could be commercially exploitable if techniques were available to recover them. A number of design possibilities exist towards this end, including continuous dragline bucket systems (Gauthier & Marvaldi 1975) as proposed for deep-sea mining operations, suction dredging systems with submerged pump or air lift as discussed for marine aggregates, systemized clamshell grabs and remotely controlled submersible units for dredging, screening and pumping the feed by flexible pipeline to a surface treatment plant. In common with future trends for marine aggregate dredging, technical feasibility is not the only consideration and the incentives for the tin industry to proceed along these lines will also depend on economic factors, costs in relation to tin prices and future world demand. Continuing increase in the price of tin could, in the long term, lead to the development of alternative substitute materials, at least for some applications, and thus become a disincentive for future developments in offshore tin operations.

Manganese nodules

Smale-Adams & Jackson (1978, this volume) have already dealt with various aspects and problems of manganese nodule mining, lifting, etc., and it remains for this paper to deal with the operational effects of the marine environment on processing and handling these materials at sea.

(a) Local weather and sea conditions

Local weather and sea conditions, together with their seasonal variations, are highly significant factors which have to be taken into account in the planning and selection of workships, floating structures and associated on-board equipment for processing or handling operations at sea. These factors must be considered with respect to stability requirements and accepted codes of practice for maritime safety as well as the technical and economic viability of the operations which are being planned.

In order to give an impression of the severity and range of conditions likely to be encountered, seasonal wind and wave characteristics for the area centred around 10° N, 140° W are given (Meteorological Office 1977). This area is one in which nodule mining is likely to take place.

(i) Wind

The NE trade winds affect the location from October to July, but during August and September winds are variable and mostly light. Mean monthly speeds vary from 8 knots in August to 17 knots in April with an annual average of 14 knots. Calms are rare as are winds above force 7; however, it is estimated that between July and September the location is affected by tropical storms or hurricanes two years out of three, when, of course, much more severe conditions would be experienced.

(ii) Waves

Swell waves are generally larger than wind waves, with a mean height of about 4 m and periods of between 10 and 16 s (wavelengths 150–400 m). During the tropical storm season in particular, swell is often heavy and confused. Significant heights of wind waves mainly vary between $1\frac{1}{2}$ and 3 m but periods are very variable, mostly 5–9 s with 6 or 7 s most common, corresponding to wavelengths of 40–125 m with a mode of about 60 m.

It is not proposed to extrapolate this information in terms of actual ship-board motion since this depends on a wide range of other factors beyond the expertise of the author. However, some comparative indication of ship's motion may be obtained by reference to figure 3. These data were recorded on a sand-and-gravel dredger with principal dimensions: length 79.2 m (260 ft o.a.l.), beam 14.02 m (46 ft) and draught 4.57 m (15 ft) at a position close to the stern of the vessel and on the centre line of the main deck. It will be seen that the mean wind speed conditions were 13.3 knots during this test as compared with the annual average of 14 knots in (i) above. Thus the out-to-out motion recorded on the dredger for heave was 2.3 m, for roll 2.4° and for pitch 4.9°. (The maximum and minimum values for these motions are computed about the mean.)

During these trials, the dredger was under the command of an experienced master, who handled the vessel to minimize motion in order to facilitate continuous and effective dredging. It is likely that ship motion, in this particular area of the Pacific discussed, will be of no less magnitude.

For mining operations, the problems of safety of the lifting systems must be a matter of some concern during periods of tropical storms and hurricane conditions which are predicted.

(b) Ship-board processing operations

Since the proposed mining sites are over 2000 km distant from land, it is important to consider what processing operations can be carried out on site in order to reject as much waste material as possible and thus to avoid the cost of transporting it long distances. However, work to date indicates that the primary operations of dewatering and removing the ocean bed sediment (ooze) are the only possibilities at this stage. In this respect, the effect of ship-board motion may interfere with classification procedures designed to reject the ooze from fine nodular particles at the lower end of the particle size range. Such effects will depend on the size distribution of the feed to any separation process, the concentration of solids and other factors, including the frequency and amplitude of the ship movement during the operations. The particle size of the nodular material and the percentage of associated ooze will depend, to a large extent, on its previous history and location, friability of the nodules, methods of harvesting and lifting, and any other handling or transfer operations. To maximize the recovery of nodular material in any particular mining operation, it may well be necessary to determine the hydrodynamic behaviour of the feed material, its response to classification processes, and how these are affected by superimposed shipboard motions of varying degrees of severity. Such data, in conjunction with the corresponding mineral distribution with particle size, will have to be used as a basis for selecting appropriate classifying operations and for predicting expected scale-up performance.

(c) Materials handling

It is conceivable that ship-to-ship transfer systems will be employed where dewatered solids have to be transferred from the mining and processing vessel to a bulk cargo vessel for transportation to the land-based processing plant. Transfer systems which have been developed (G.E.C. Mechanical Handling Ltd 1973) and used for many years in naval and mercantile replenishment-at-sea applications, could readily be adapted for this purpose. For example, such systems incorporate methods for refuelling using self-sealing dry couplings, transfer of stores and ammunition, and the safe transfer of personnel between vessels under way. Hydraulically driven winches and solid-state logic control systems provide the necessary sensitivity and flexibility required for safe operation when two ships, often of very different tonnages, are keeping stations when underway in quite severe sea states.

Similar equipment and methods have been relied on extensively for many years on self-unloading bulk cargo vessels where no shore handling equipment is available.

There seems little doubt that similar systems could be used for ship-to-ship transfer of nodular material either as hydraulic suspensions or partially dewatered bulk solids.

(b) Transportation of damp nodule cargoes

The transportation of bulk mineral concentrates and other similar fine-grained materials requires their carriage in a manner such as will avoid the hazard of shifting cargo due to liquefaction. The onset of the phenomenon is associated with the moisture content exceeding some critical limit in relation to the prevailing conditions. Such cargoes may appear to be in a relatively dry granular state when loaded and yet may contain sufficient moisture to develop a flow state under the stimulus of vibration and ship's motion, particularly during a voyage in heavy seas. In the resulting fluid state, a shift of cargo can occur in various different ways, from which the vessel may progressively reach a dangerous heel and eventually capsize.

There have been a number of casualties attributed to this cause over the years, in some cases involving loss of life (Green & Hughes 1977). In others, shift of cargo with consequent development of a list condition have been reported but the vessels succeeded in safely reaching port.

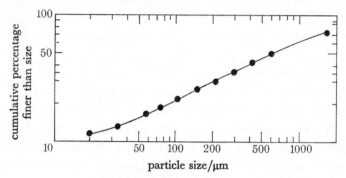

FIGURE 8. Size analysis of anthracite washed duff.

Any damp fine-grained mineral cargo – and this must include damp or wet fine-grained nodular material – is regarded as lying within the category of potentially hazardous material for shipment as a bulk cargo. It should therefore be subjected to internationally recognized flow moisture tests from which its transportable moisture limit for safe transport by sea can be derived (IMCO 1972). Furthermore, since there is no background of experience in the shipment of this material, as, for example, exists for ore concentrates and fine coal, confirmation of its flow moisture characteristics and behaviour under simulated shipboard conditions is advisable.

The effect of simulated ship's motion and vibration are dramatically illustrated in figure 7, plate 2. This shows a model-scale cargo container in which an anthracite washed duff at a moisture content of approximately 21 % is being tested on a ship's motion simulator. The particle size of this sample is shown in figure 8.

It will be seen that even though the cargo appeared reasonably dry when loaded, the onset of liquefaction occurred near the top of the ridge within 3 min and gradually transformed the cargo into a fluid mass inside 1 h. Under real conditions on a bulk cargo vessel this could create a situation of extremely hazardous instability.

Although various test procedures and codes of practice have been evolved with a view to ensuring that such cargoes have sufficiently low moisture content to be transported safely by sea, cargo behaviour during a voyage is far from being fully understood. In this situation it cannot therefore be assumed that because manganese module material in a partially dewatered state may appear dry and safe, it is actually safe to transport as a bulk cargo by sea. The alternative is specially designed ships with compartmented cargo holds to reduce the free surface problems, as in tankers. However, this increases capital and the materials-handling cost and must be considered in relation to dewatering costs to bring the cargo within the transportable moisture limit as a bulk cargo.

Red sea mud

The locations of brine pools and metalliferous sediments along the median valley of the Red Sea have been investigated by different groups of scientists since 1964. Their location is shown in figure 9.

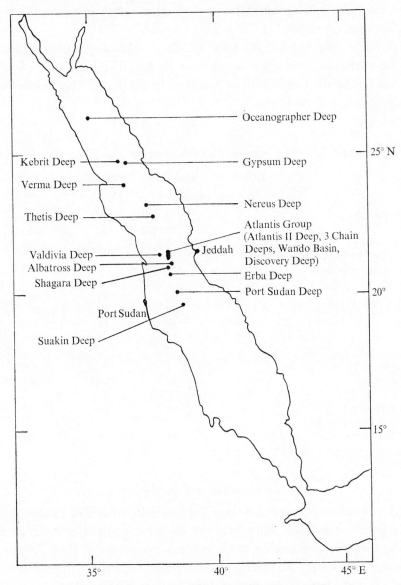

FIGURE 9. Distribution of brine pools in the Red Sea.

The largest of these pools is the Atlantis II Deep which covers an area of 55 km² and contains the only sedimentary deposits at present of economic interest in the Red Sea. The southern part of Atlantis II Deep (Amann, Bacher & Blissenbach 1973) is estimated to contain 150–200 Mt of dry salt-free mud with an average content of 5 % Zn, 1 % Cu and 33 % Fe (Supp & Nebe 1974). These metals occur mainly as sphalerite, chalcopyrite and iron oxides or hydrates of iron oxide respectively. In the original state, one litre of thermal sediment contains about 90 g of dry salt-free solids and 300 g of salt (99 % NaCl). The deposits are located at a depth of about 2 km in a layer up to 10 m in depth with temperatures up to 60 °C. The solids are characterized by their extremely fine particle size, of the order of 60–70 % by mass finer than 5 μm. This creates considerable problems in developing technical and economically viable processes for extracting the metal values. However, investigations to date appear to indicate that froth flotation (Supp & Nebe 1974) followed by hydrometallurgical treatment (Neuschütz & Scheffler 1977), leaching, solvent extraction and electrowinning processes are possible routes for recovering the mineral values.

In this event and bearing in mind the bulk handling and transportation problems, it seems reasonable to speculate that primary concentration by flotation could take place on site with subsequent treatment of the concentrates on a land based plant. The former would involve processing under shipboard conditions.

TABLE 1. WIND AND SEA CONDITIONS IN THE RED SEA

	Dec.–Feb.	Mar.–May	June–Aug.	Sept.–Nov.	year
(i) winds					
predominant direction(s)	N	N (NW)	NW	NW (N)	—
frequency (%)	47	40 (30)	45	33 (30)	—
mean speed (knots)	10.3	10.0	8.3	8.0	9.2
hourly mean speeds expected on average to be exceeded only once in:					
1 year	34	35	31	31	37
5 years	37	39	34	34	41
10 years	38	40	35	36	42
50 years	41	43	38	39	45
(ii) waves					
height ⩽ 2 m (%)	84	81	93	90	87
period ⩽ 7 s (wavelength ⩽ 75 m) (%)	77	75	80	77	77
maximum wave exceeded only once in 50 years/m	10	11	8	9	12

(a) Local weather and sea conditions

The same general principles apply as those expressed in an earlier section for mining and processing of manganese nodules. However, for the area centred around 21–22° N, 38° E in the Red Sea, there is considerably more information available than for the area where manganese nodules are located. An attempt has therefore been made to predict extreme winds and waves and the data have been divided into four periods roughly corresponding to the northern hemisphere seasons. The results thus obtained are given in table 1 (Meteorological Office 1977).

No specific information is available on swell at the location of Atlantis II Deep but the relatively small area of the Red Sea suggests that this is not a significant factor.

It will be seen that conditions for mining and processing operations in the Red Sea are likely to be more constant and less severe than those predicted for corresponding operations on manganese nodules in the North Pacific.

(b) Mining and shipboard processing operations

Mining methods for Red Sea muds are likely to be relatively simple compared with those for harvesting manganese nodules. First, the operating depth is of the order of 2 km as compared with 5 km for nodules. Secondly, the deposit is in the form of a deep layer of sediment about 10 m thick as compared with relatively thin layers of nodules deposited on the seabed. Consequently, in principle, much simpler methods can be adopted for lifting the material to the surface, such, for example, as suction dredging. However, materials of construction for both mining and surface treatment plant are much more significant factors in view of the high temperature (60 °C) *in situ* and brine concentration (300 g l^{-1}).

In decisions regarding processing on site, the same general principles apply as for manganese nodules. Similarly, the motion produced by local sea and weather conditions will also interfere with the separation behaviour of the particles which in this case is likely to be by froth flotation. Briefly, this involves selective preconditioning of the particles with surface active reagents in an aqueous slurry and agitation of the pulp, which at the same time induces air as finely dispersed bubbles. The addition of a frothing agent enables the fine mineral particles which rise to the surface attached to the bubbles to be skimmed off by means of a rotating paddle.

Shipboard motion in varying degrees of severity will produce a number of effects on these processes:

(i) Angular motion will reduce the effective volume of the flotation cell due to overspill as the cell is tilted towards the discharge lip. Thus the throughput on a commercial scale will also be reduced for a given size of plant.

(ii) Removal of froth is likely to be less uniform due to the flow of froth towards or away from the discharge lip as illustrated in figure 10, plate 3.

(iii) Directional orientation of the discharge flow should therefore be in the same direction as that of the minimum motion in the prevailing sea state.

(iv) Increased froth volumes can adversely affect the grade and recovery of mineral values in the concentrate.

(v) The speed of the paddle should ideally be arranged to avoid coincident phase relations with the frequency of the predominant wave motion.

(vi) The interaction of heave motion and slamming may also affect performance.

(c) Materials handling and transportation

The same general principles and requirements apply as discussed for manganese nodules. However, if a concentration step is carried out at the mining site, the transportation operations will be on a much smaller scale.

PRESENT AND FUTURE RESEARCH

In order to investigate the effects of motion on the various processing and other operations likely to be carried out at sea, the first requirements, as part of a research programme at Warren Spring Laboratory, were clearly to have the ability to monitor the sea-board motion of dredging, mining and other types of workships over the full range of operating conditions and then to reproduce them on land-based ship's motion simulators. Thus it would be possible to conduct investigations on how equipment and process behaviour was affected, not only over a range of conditions of varying severity, but also by individual motions, e.g. roll, pitch, heave, etc.

(a) Monitoring and simulation of a ship's motion

Successive sea trials have led to the progressive development of ship's motion monitoring equipment to meet these special requirements. In addition, a recently developed system allows for submersible applications, providing similar outputs to a depth of 200 m at a maximum cable length, at present, of 400 m. In both these systems, the results produced can be immediately displayed on board ship as well as being recorded for detailed examination when the sea trials have been completed.

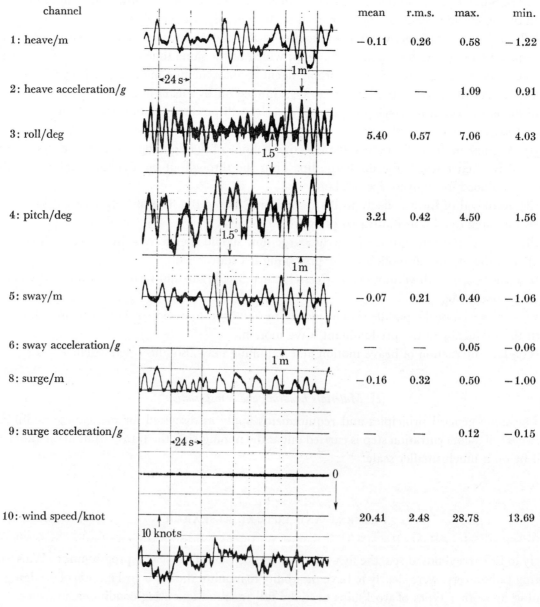

channel	mean	r.m.s.	max.	min.
1: heave/m	−0.11	0.26	0.58	−1.22
2: heave acceleration/g	—	—	1.09	0.91
3: roll/deg	5.40	0.57	7.06	4.03
4: pitch/deg	3.21	0.42	4.50	1.56
5: sway/m	−0.07	0.21	0.40	−1.06
6: sway acceleration/g	—	—	0.05	−0.06
8: surge/m	−0.16	0.32	0.50	−1.00
9: surge acceleration/g	—	—	0.12	−0.15
10: wind speed/knot	20.41	2.48	28.78	13.69

FIGURE 11. Typical set of recordings taken on board a drillship.

Measurements are made of roll, pitch, heave, sway and surge. These, together with wind speed, direction and any commentary that is required, are recorded on magnetic tape. A typical set of recordings taken on board a drillship are shown in figure 11.

Three ship's motion simulators, also developed at various stages of the programme to meet particular requirements, are now available for equipment and process testing. The largest and most recent one, which has a heave displacement of 3.7 m, with a payload of 1.5 t, as well as roll and pitch motions, is illustrated in figure 12, plate 4. Both facilities are described in greater detail elsewhere (Warren Spring Laboratory 1977), but their principal specifications are given in table 2.

TABLE 2. SHIP'S MOTION MONITORING AND SIMULATION FACILITIES

Description of facility

ship's motion monitoring systems
    capability of monitoring and recording angles of roll and pitch; vertical, lateral, fore-and-aft accelerations and
      displacements, also vibration
    basic specification:
      roll and pitch range of $\pm 25°$ to an accuracy of $\pm 1\%$;
      heave range of ± 15 m to an accuracy of $\pm 5\%$;
      sway range of ± 15 m to an accuracy of $\pm 10\%$
      (all of which can be measured over the frequency range of 5–25 s)
      vibration can also be measured.

ship's motion simulation facilities
    capability:
    1. (a) Up to 1.2 m heave
       (b) Up to 15° out-to-out angles of roll and pitch
       (c) Payload 200 kg
    2. (a) Up to 1 m heave
       (b) Up to 25° out-to-out angles of roll
       (c) Up to 12° out-to-out angles of pitch
       (d) Up to 6° out-to-out angles of yaw
       (e) Payload 1 t
    3. (a) Up to 3.7 m heave
       (b) Up to 16° out-to-out angles of roll
       (c) Up to 16° out-to-out angles of pitch
       (d) Payload 1.5 t

(b) Research applications

Up to the present, a wide range of investigations have been carried out in relation to the exploitation and transportation of minerals from the continental shelf and from the deep seas, including various studies on mineral processes and equipment such as: sedimentation and flotation; feed and distribution systems, spiral and pinched sluice concentrators, mineral jigs; liquefaction phenomena and the behaviour of fine-grained bulk cargoes at sea; instrumentation and other equipment for the offshore oil industry.

(c) Future research requirements

Various problem areas and future requirements have been discussed in the preceding sections. However, in the present climate of confidentiality by the different commercial organizations involved in both continental shelf and deep-sea mining activities, it is difficult to specifically state what has been done, what is being done, or what still needs to be done. At best, from what may be regarded as common knowledge, one can only speculate on where the problem areas are likely to be.

An attempt has been made to summarize some of these in the following list:

(i) Dredging systems for deeper water on the continental shelf. (For the deep oceans this is mainly in the hands of the various consortia involved.)

(ii) Primary underwater processing systems, size classification, gravity separation and possibly comminution if commercially exploitable hard rock deposits are discovered.

(iii) Materials of construction and design systems to withstand salinity, pressure and temperature of the deep sea environment in mechanical, electrical and control applications.

(iv) Evaluation of the effect of shipboard motion – as arising (ongoing at W.S.L.).

(v) Improved compensating systems or alternatively semisubmersibles or submersibles.

(vi) Specially designed mineral processing equipment for operation at sea such as mineral jigs, spiral concentrators and flotation cells (ongoing at W.S.L.).

(vii) Adaptation of ship-to-ship transfer systems for materials handling, as required.

(viii) Further investigations of the factors governing the behaviour of fine-grained bulk cargoes at sea (ongoing at W.S.L.).

(ix) Disposal of waste products and their effect on the environment and ecology.

REFERENCES (Hughes)

Amann, H., Bacher, H. & Blissenbach, E. 1973 In *Fifth Offshore Technology Conference*, Houston, Texas, vol. 1, pp. 345–358.
Anon. 1977 *Min. J.* **288**, 141.
Beckmann, W. C. 1975 In *Oceanology International 75 Conference*, Brighton, pp. 342–345.
Department of the Environment 1975 *Aggregates: the way ahead*. Report of the Advisory Committee on Aggregates.
G.E.C. Mechanical Handling Ltd 1973 *Marine Equipment*. Publication no. E/Mar/505.
Gauthier, M. A. & Marvaldi, J. H. 1975 In *Oceanology International 75 Conference*, Brighton, pp. 346–349.
Granville, A. 1974 *Prospects for the recovery of minerals from the sea* (NIM Report N 1644). Nat. Inst. Metall. Johannesburg, R.S.A.
Green, P. V. & Hughes, T. H. 1977 *Trans. Instn Min. Metall.* A **86**, 150–158.
Hewitt, J. A. 1977 F. W. Payne & Son (Bickley) Ltd, Kent. (Private communication.)
Hughes, T. H. & Joy, A. S. 1972 In *Oceanology International 72 Conference*, Brighton, pp. 303–310.
International Tin Council 1975 *International Tin Council statistics*.
Inter-Governmental Maritime Consultative Organization (IMCO) 1972 *Code of safe practice for bulk cargoes*.
Institute of Geological Sciences 1977 Mineral Statistics and Economics Unit, London. (Private communication.)
Meteorological Office 1977 Met.03C, Bracknell, Berks. (Private communication.)
Neuschütz, D. & Scheffler, U. 1977 *Erzmetall.* **30**, 152–157.
Smale-Adams, K. B. & Jackson, G. O. 1978 *Phil. Trans. R. Soc. Lond.* A **290**, 125–133. (This volume).
Supp, A. Ch. & Nebe, R. 1974 *Erzmetall.* **27**, 321–329.
Warren Spring Laboratory 1977 *Not all at sea* (brochure of film). Central Office of Information Central Film Library, Cat. No. UK 3317, London.

Discussion

B. WHITE. (*Department of Mineral Resources Engineering, Royal School of Mines, Prince Consort Road, London, S.W.7*). I should like to ask Mr Hughes to stick his neck out and make some prediction of the future of marine mining processing systems.

Although it may be easier going into deeper waters when considering oil-pollution problems (see Dr Gaskell's paper, which follows), it most certainly is not when considering mineral processing. We already have the problem of developing new technology for the mining of the nodules, etc., without even considering their processing at sea. Is there not a good case for compensating the operational environment and making use of existing processing technology rather than redesigning all the equipment?

T. H. HUGHES. The design of any mineral processing plant is invariably based on previous feasibility studies of the composition, characteristics and behaviour of the constituent minerals in the ore body which has to be processed. However, in addition to the prime technological considerations, a number of other important factors such as capital and operating costs, amortization period, location and site conditions, availability of supporting resources, etc., have to be taken into account before final decisions on the design and selection of the plant are reached.

In general, I think the same broad principles would apply in the design of a concentrating plant for the exploitation of a marine mineral deposit. The use of a compensating system or alternatively redesigned equipment for a particular deposit would be an additional factor for consideration with all the others. Certainly, technology exists for compensating systems or could readily be developed as, for example, from the advanced designs for gun or rocket launching platforms used at sea, to very simple systems tested at W.S.L. (Hughes 1973) for pinched sluice concentrators and subsequently for spiral concentrators. Such systems would need to be evaluated during the feasibility studies.

Therefore, at present it is not possible to advocate one method of approach or the other. However, further studies of the effects of ship-board motion on the various processing operations, leading to new designs of equipment which will operate more effectively in a marine environment than existing types, would enable decisions on final plant design to be reached more conclusively.

Reference

Hughes, T. H.   1973   In *Proc. Joint Technical Meeting of the Society for Underwater Technology and the Institution of Mining and Metallurgy*. Warren Spring Laboratory, Stevenage, Herts.

Phil. Trans. R. Soc. Lond. A. **290**, 179–185 (1978) [179]

Printed in Great Britain

Environmental pollution in offshore operations

By T. F. GASKELL

Oil Industry International Exploration and Production Forum,
37 Duke Street, St James's, London SW1 6DH, U.K.

The oil industry has made contingency plans from the earliest operations to clean up any spills due to accidents during exploration and production. The Oil Industry International Exploration and Production Forum (E & P Forum) was established in 1974 to coordinate oil industry opinion with governments and intergovernmental agencies. The United Nations Environment Programme and the Intergovernmental Maritime Consultative Organization are working together to organize regional oil spill clean-up arrangements both on the apparatus and on the financial side.

The E & P Forum member companies have formulated a computer program to estimate the cost of clean-up following any hypothetical blow-out in the North Sea. This program is applicable, provided the appropriate meteorological data are available, to other areas of the world such as the Mediterranean, the Persian/Arabian Gulf, the Gulf of Mexico, the Malacca Straits, etc., where U.N.E.P./I.M.C.O. are proposing to set up multi-national conventions.

In places such as the North Sea, the oil companies have mutual aid organizations which maintain stocks of dispersant, suitable vessels and spreaders, skimmers and booms to minimize any oil that may be driven to the shore.

Looking to the future, there may be some problems associated with harvesting manganese nodules or mining the Red Sea mineral-rich lands. However, conservation is today part of any general planning of an operation and new processes that are developed will take into consideration adequate anti-pollution measures.

INTRODUCTION

The record of the offshore oil production industry in environmental pollution is in general a good one. The most active area in the world has been the Gulf of Mexico, where offshore oil production started in earnest in the 1940s, and where, for example, in 1972, 17 % of the U.S.A.'s domestic crude oil originated. Although some mistakes have been made and accidental blow-outs have occurred, there is no obvious ecological damage along the Louisiana and Texas coasts. Offshore exploration and production is now taking place in many of the continental shelves of the world and the lessons learned from the earlier activities are being applied, with suitable climatic and political modifications to new areas.

The volume of the oceans is 300 million cubic miles (*ca.* 1250×10^6 km^3); the total world oil production is one cubic mile (*ca.* 4.2 km^3) per year so that, provided sufficient mixing takes place, the small percentage of oil that is spilt becomes sufficiently diluted for it to get lost by natural processes. Since oil is formed in a marine environment from the decay of animal and plant matter, there are always hydrocarbons of a petroleum-like nature associated with recent sediments on the sea bed, and there do not appear to be any animal or plant chains of life that concentrate hydrocarbons as is the case in some instances with heavy metals.

In looking ahead at offshore mineral production, there should not be any oil problems that have not already been experienced. The possibilities for minerals such as manganese or phosphatic nodules causing deleterious effects in the vast volume of the oceans are unlikely, since they already exist in contact with the sea and the ocean floor and they are therefore in as strong a concentration locally as they would be in any extraction process.

OIL SPILL STATISTICS

Several surveys of the main sources of oil in the sea that have been made during the past few years (National Academy of Sciences 1974) agree that the offshore oil drilling and production side of the industry is responsible for only a comparatively small amount of oil pollution that is less than 2% or 30000 tons per year in a total of about 2 million tons a year. The accepted make-up of oil in the sea is shown in table 1.

TABLE 1

	quantity/(t/a)
ballast and wash water from tankers:	
(a) non load-on-top	670000
(b) load-on-top	120000
machinery space bilges	24000
accidents to ships:	
(a) tankers	135000
(b) other vessels	23000
effluents from shore	500000
offshore oil drilling and production	34000
natural sources (seepages underwater)	500000

Although incidents such as the *Torrey Canyon* wreck which spilt over 100000 t of crude oil near shore are notorious and heard of world-wide, only one such accident (or an equal tonnage of smaller ones) occurs on average each year. Four times as much waste oil goes down with the untreated sewage from hundreds of large coastal cities who cannot afford modern plant to clear their effluent, and who prefer to put it straight into the sea.

The offshore oil contribution to total oil spill is less than a quarter of that from tanker accidents, and although more wells are coming into production each year, the accident rate does not appear to be increasing. This is because each accident provides lessons learned and consequent changes in codes of practice and improvements in safety and fail-safe devices.

The figure for oil introduced into the sea from natural seepage is possibly much greater than given above since only those seepages on or close to the shore have been identified. In cases where seepages have existed throughout historical times there is no observable bad effect on the fauna and flora. Oil spills in the sea are probably more a nuisance value to human beings than a harmful agent to other animals.

LOUISIANA EXPERIENCE

Evidence concerning the limited environmental effect of oil operations has been provided by a recent comprehensive study by the Gulf Universities Research Consortium. The objective study, known as the Offshore Ecology Investigation or O.E.I., was to determine the effect of petroleum operations in the 'Louisiana oil patch', the continental shelf area which has experienced about 25 years of intensive petroleum exploration and production operations. If, indeed, exploration and production operations have an adverse effect on the ecosystem, here is where one might expect the evidence of such effects. This is particularly true with regard to long-range effects which might be hidden by natural phenomena over the short term. That study indicated the following:

1. Natural phenomena completely dominate the characteristics, productivity and general health of the ecosystem. These include seasonal changes in water quality, water mass movement, and the turbid layer arising from the Mississippi River, which contributes far more to silting and sedimentation than does production and drilling activity.

2. The presence of offshore producing platforms and pipelines has an insignificant effect which, if anything, appears beneficial, owing to the reef effect of the structures increasing the productivity of basic nutrients in the vicinity.

3. Petroleum operations have not resulted in any significant accumulation of potentially toxic materials in either the sediment or water column in the vicinity of such operations.

4. No accumulation of hydrocarbons was found in the animal life in the area and the accumulation of organic materials in the sediments and beach sand was found to be of a low order and not ecologically significant.

The report also emphasized that in so far as environmental protection is concerned, there was little comparison between the exploration and production operations of the late 1940s and 1950s in shallow waters (which would presumably be more susceptible to ecological damage) and operations today with advanced technology in deeper waters more remote from the coast.

THE E & P FORUM SUBCOMMITTEES

The Oil Industry International Exploration and Production Forum (E & P Forum) was established in 1974 as an association of oil explorers and producers in order that the industry may speak with a united voice to international agencies who are interested in formulating regulations to control oil activities. The membership is of three types: oil companies engaged in finding and producing oil, national oil companies, and regional groups such as the U.K. Offshore Operators Association, the American Petroleum Institute, the North Sea Operators Committee in Norway. The E & P Forum has its office in London, since this is where the Intergovernmental Maritime Consultative Organization (I.M.C.O.) of the United Nations is situated. I.M.C.O. has in the past dealt with international aspects of safety and oil spillage from tankers and have now turned their attention to offshore oil operations. The E & P Forum has been awarded non-governmental observer status with I.M.C.O. and assists with technical advice on relevant drilling and production matters.

The E & P Forum is a member of the International Petroleum Industry Environmental Conservation Association (I.P.I.E.C.A.) and has assisted on the exploration and production side of the oil business in meetings in Tehran (1975) and Paris (1977) to discuss with the United Nations Environment Programme (U.N.E.P.) the impact of petroleum on the environment.

Subcommittee A of the Forum deals with pipelines and a yearly survey is made of any accidents to pipelines in the North Sea and the Persian/Arabian Gulf. This survey will be extended to new production areas of the world with a view to continual improvement of techniques from lessons learnt in practical operations.

Subcommittee B deals with I.M.C.O. affairs and arranges for appropriate experts to attend I.M.C.O. meetings to advise on technical matters, after gathering the concerted opinion of the industry through working groups drawn from E & P membership.

Subcommittee C, on oceanography and meteorology, helps to coordinate the information on climate and weather which is needed for successful operations at sea. In the first instance, accurate forecasting of waves and wind is needed for critical operations such as pipe-laying or moving large equipment from shore to operational site. An extension of this day-to-day forecasting is the average weather picture needed in order to plan ahead and cost the effect of unavoidable delays. In addition to forecasts, the long-term wave current and tidal data are needed to allow adequate and economic design of offshore structures which will stand up safely to the prevailing conditions. The exchange of information, for example, between Norway and Britain, on conditions and guidelines produced by governments is important where design criteria are laid down by national regulations. The experience gained in one region of the world enables subcommittee C to advise operators in new places on the type of measurements that will be needed for both design and operational purposes.

Subcommittee E looks after the long-term design of offshore structures, and maintains liaison with the I.M.C.O. committees on stability, etc., and also with subcommittee C on matters pertaining to wave and wind forces on structures at sea.

Subcommittee D looks after the legal aspects of offshore operations and was formed to keep liaison with the Offshore Pollution Liability Association Ltd (Opol) and to help to extend this arrangement from the U.K. sector of the North Sea to other parts of the world. The Opol argument has been extended to cover most of the active area of the North Sea, but may be superseded by the Nine Nations Convention on Civil Liability in the North Sea. This convention at their meeting in London appointed the E & P Forum to be their technical advisers, and the Forum members have produced a computer program to simulate the movements and quantities of spills consequent on accidents in present North Sea oilfields, together with the cost of clean-up operations and awards for damages. Although in the early stages some astronomical costs for single incidents were proposed by some nations, the technical aspects were considered and some form of general limit of about $30 M per incident were agreed for civil liability, although this agreement was to some extent vitiated by a loophole clause which allowed individual nations to assess the industry with unlimited liability. It was hoped that the nine nation convention would provide a workable pattern for other regions of the world where offshore production may take place, but it appears to have confused the issue rather than to have provided a universal workable arrangement between governments, oil producers and the insurance world.

Subcommittee F was formed to provide technical advice to subcommittee D on all aspects of oil pollution prevention in offshore operations.

THE SLIKTRAK COMPUTER PROGRAM

During discussions on civil liability it became apparent that there was a need to illustrate effectively and realistically the combined efforts of clean-up activities and natural spill-reducing phenomena, and so predict the quantities of oil that could reach the shore. Estimates were made of the amount of oil that might be emitted in a platform blow-out and the duration of the emission before the wells were brought under control. The known rate of evaporation from the surface film of oil and estimated amounts of oil that could be contained and picked up by booms, skimmers, etc., were fed into the analysis. The rate of speed of the oil slick is determined by the wind speed and for this purpose the computer selected random weather from determined meteorological data over the past three years. The weather conditions which, if severe, could hamper containment operations, would increase the loss of oil slick by natural mixing with the sea water. The removal of oil from the surface in this way was included in the program based on rates of mixing determined by the U.K. Government research station at Warren Springs. The cost of breaking up oil slicks with dispersant was included in the cost of the operation, and the remaining slick, driven mainly by the wind, was followed until it disappeared completely or reached the shore. The effect of currents was included, but is very small compared with wind effect in the North Sea, especially as most North Sea water movement is caused by tidal streams and reverses with the tidal cycle. The computer assessed the cost to fishermen of temporary loss of livelihood, and to hotel and boarding house keepers of diminishing of tourist trade as well as the local authorities and government bills for cleaning beaches. Five thousand random incidents were analysed by the computer on this basis. The simulation showed that the average total spill cost is expected to be $6 M, that the probability is 90 % that the spill cost will not exceed $15.8 M and that the worst case would amount to less than $25 M in civil liability.

Although the SLIKTRAK program (Blaikley *et al.* 1977) is designed for random accidents to production wells in the North Sea U.K. sector, it can be readily adapted to other areas, and with small modification to replaying real incidents with actual weather data.

The recent *Ekofisk Bravo* incident in the North Sea, where the total oil spilt was less than a quarter of the *Torrey Canyon* spill, and where the point of spill was more than 100 miles from shore, shows that the SLIKTRAK prognostications are on the right lines and if anything overrate the civil liability costs. The *Bravo* spill was, like accidents in the airline industry, due to a combination of human errors and fallibilities, and was put right by the opposite facet of human abilities: bravery, ingenuity and experience in adversity.

The computer program provides not only a useful yardstick for legislation in other parts of the world, where extensive offshore oil production may develop, but also shows on what beaches spills are likely to arrive, and the size of the clean-up problem that may be posed. It is probable that more experimental figures for evaporation and mixing of oil with water will be required for different climates, but, provided the meteorological data are available, the general lines of thought of the SLIKTRAK program are applicable.

NORTH SEA CONTINGENCY PLAN

The U.K. offshore operators have for many years had contingency plans for coping with accidental oil spills during exploration and production operations. Stocks of dispersant are maintained at strategic points and the supply lines to the manufacturing sources are arranged. The devices needed to spread the dispersant are held ready for fitting to vessels which are on call from their normal duties when an emergency arises. The locations of company-owned booms and oil collectors are known to all operators and a chain of command and liaison with government authorities is laid down.

It was notable in the *Ekofisk Bravo* spill that the contingency plans worked well; all the other producing wells on the platform shut down automatically; the 120 man crew of the platform were evacuated without injury; rigs for drilling relief wells were made available; and fire fighting vessels were soon on the scene.

Similar mutual self-help arrangements have been made for the Gulf of Mexico and the Persian/Arabian Gulf, and there is no doubt that the oil offshore production industry is prepared for emergency. Recently I.M.C.O. and the Intergovernmental Oceanic Commission (I.O.C.) have, with the financial support of U.N.E.P., formulated regional clean-up schemes for various areas of the world. A monitoring centre has been set up in Malta to determine the size of the pollution problem in the Mediterranean. One of the substances being monitored is oil. Similar regional schemes are proposed for West Africa, the Persian/Arabian Gulf, the Malacca Straits and Southeast Asian waters, and the Caribbean. In those places where oil production is well established, oil industry arrangements will be available as a basis for inter-governmental cooperation.

DEEPER WATER PROBLEMS: OIL

In most parts of the world, moving into deeper water means that production will be farther from the shore. The chance of oil spills reaching land will therefore be reduced, so that there is no reason why the movement of the oil industry into deeper water should provide any greater problem of pollution (Blaikley 1977). There may be greater difficulty in reaching the source of spill for relief vessels, but this effect will be marginal, especially when unforeseen delays can occur in any case due to bad weather.

There will of course be a much greater opportunity for the oil to mix with the ocean water, except in the case of some crude oils, which seem to form 'tar-balls' which can persist with an outer coating of weathered oil for many months. However, these persistent blobs of oil are generally formed from the heavy sludge from the bottom of fuel oil tanks rather than from crude oil with its usual mixture of a wide range of hydrocarbons.

DEEPER WATER PROBLEMS: MINERALS

Any extraction of mineral wealth from the seabed, either by mining manganese nodules or sucking up mineral rich brines from the Red Sea, will take place in deep water, which means that there will be ample opportunity for dilution of any waste products. These may be clouds of dust particles produced when manganese nodules are washed or when unwanted mud and sand are put back into the sea. These clouds of dust may be carried for miles before they gradually descend to the seabed. This process takes place all the time when rivers bring down sediments derived from the land, and there is no conceivable way in which this material can pollute the varigated seabed picture that nature produces. There may be some confusion in the meaning of the geological evidence from cores taken in the seabed in years to come, but it will be a similar trouble for the marine geologist as has been provided by the carriage of erratic materials by icebergs.

New industry is fortunate compared with old established extraction processes such as those for coal and oil, which grew up in a small way before human beings became conscious of the nuisance and danger of pollution in the seas. New processors are well aware of the need to make impact surveys to forecast what effect their operations will have on the environment, and they can include necessary precautions in their planning. The older industries have had to face the more expensive problem of making modifications to their previously accepted techniques. In all cases, however, industry realizes what has to be done and has engineers and scientists as well qualified as those in other walks of life to study the problems and their solutions.

Addendum (21 April 1978)

It is interesting to add a footnote concerning the tanker accident from the ground of the *Amoco Cadiz* on 17 March 1978. This was a very large tanker containing 200 000 t of oil, but the result of the accident will be seen eventually to bear out what has been stated above. The tanker, of course, had nothing to do with exploration and production which are the subject of this paper. The spill accounted for less than twice the annual average for tanker accidents, which as all those versed in probability studies will agree is hardly out of the ordinary. The long-term effect on marine life will be found to be negligible. It is already reported that most of the oysters were not killed, and it is thoroughly well demonstrated by marine biological experiments that oil assimilated by marine life is cleaned out in two weeks sojourn in clear water. It is probable that a small criticism could have been made of the authorities because they did not start where the *Torrey Canyon* experience of 10 years ago left off, but there were extenuating political conditions of a current election. A great deal is known about the best way to cope with accidental oil spillages and how to clean up oil covered beaches and it is time the academic world and the media applied their normal commonsense to what is necessarily an emotive problem, that is, to use past experience to guide them into the best solution. In the long run, of course, that is in about a year, it will be rough weather and the pounding of beaches by waves that clears up the mess. Meanwhile, it should be appreciated that oil in the proper place brings benefit to man.

REFERENCES (Gaskell)

Blaikley, D. R. 1977 Environmental protection in North Sea exploration and production operations. *Marine Policy*, April. (This paper contains useful references for further reading.)

Blaikley, D. R., Dietzel, G. F. L., Glass, A. W. & van Kleef, P. J. 1977 SLIKTRAK – a computer simulation of offshore oil spills, clean-up and associated costs. *Proceedings of Oil Spill Conference*, E.P.A./A.P.I./U.S.C.G., *New Orleans, Louisiana, March* 1977.

National Academy of Sciences 1974 *The fate and effect of oil in the marine environment*. Washington: National Academy of Sciences.

Discussion

B. WHITE (*Department of Mineral Resources Engineering, Royal School of Mines, Prince Consort Road, London S.W.* 7). I should like to congratulate Dr Gaskell on his paper. It is good to hear such a reasoned and pragmatic discussion of the problems associated with the exploitation of the mineral wealth of the oceans rather than the more usual emotional outbursts of the more vociferous of the 'environmentalist' lobby.

I concur with his comments about the awareness of the potential exploiters of the deep ocean mineral resources. All the consortia involved report that work is being done to evaluate the possible environmental pollution hazards and to determine remedial measures; notable is the joint Industry/U.S. Government work on the Domes Project (Deep Ocean Mining Environmental Study). We at the Royal School of Mines are undertaking sponsored research work on the disposal of tailings following shipboard processing as part of the studies into the feasibility of exploiting the metalliferous muds of the Atlantis II Deep in the Red Sea.

INDEX

(Prepared by M. R. Strens)

Abyssal depths, ferromanganese nodules, 43, 44, 49, 55
hydrocarbon yield, 4
Abyssal plains, 35
Access, parallel, 27
Acoustic mapping, 111
Aluminium alloys, use in superstructures, 18, 19
Amoco Cadiz, 185
Anchoring problems, 110
Arcolprod, 8
Articulated columns, 9
Atlantic Ocean, deep water slope sediments, 39
margins, 37, 75, 83
shallow water slope sediments, 37
thickness of sediments, 38
'Auguste Piccard', 137, 146
Australia, NW shelf, 40

Barium, in ferromanganese deposits, 43, 46, 55, 58
Barytes, offshore mining, 161
Basement features, 6
Basins, ocean, 35
oil in sedimentary, 3
Los Angeles, 40
Bauer Deep, ferromanganese deposits, 57–64
Birnessite, *see* Manganese oxides
Blow-outs, 179
Ekofisk Bravo, 183
Brent field, economics, 99
'wet well', 102
Buoy, data, 110
underwater, 119

Calcium carbonate, offshore mining, 161
Challenger Expedition, 43
Circulation, mean, of ocean, 90, 91
Cobalt, enrichment in sea mounts, 50, 53, 54
in ferromanganese deposits, 43, 46, 49, 56, 58, 126
Conference of the Law of the Sea, 21, 23
Contiguous Zone, 21, 22
Continental crust, rifting, 75, 76
Continental margin, Arctic Ocean, 41
around British Isles, 83
Atlantic Ocean, 37–40
currents on, 87–98
hydrocarbon potential, 113
Indian Ocean, 40
morphology, 75–85
oil prospects, 36
Pacific Ocean, 41
passive, evolution, 76–79
Continental rise, 75, 76, 79
hydrocarbon potential, 113
Continental shelf, 28, 75, 76, 79
breadth, 28

convention, 22
exploitation, 28
exploration, 28
North Sea, 28
West of Scotland, 28
Continental slope, 33, 75, 76, 79, 84, 87, 93
drilling on, 99
hydrocarbon potential, 113
Contractor, underwater, 153–159
Conventions of 1958, 21
continental shelf, 22
fishing, 22
high seas, 22
territorial sea, 21
Copper, enriched nodules, 53–55, 67
in ferromanganese deposits, 43, 46, 49, 58, 126
in Red Sea mud, 171
sources of, 25
Coral reefs, as reservoir rocks, 36
gas in E Indies, 36
Current meter, 88
Currents, climatic and seasonal variability, 91
contour, sedimentation, 82
deep ocean eddies, 92, 93
erosion and deposition, 75, 79
Eulerian measurements, 87–89
inertial oscillations, 94, 95
internal waves, 96
Lagrangian techniques, 89, 90
mean circulation, 90, 91
measurement, 87–90
mesoscale variability, 92
meteorological disturbances, 93
of tidal period, 95
on continental margins and beyond, 87–98
profiling, 95
surface waves, 97
variability, 90–97

Deep Sea Drilling Project, 4, 5, 35, 40, 113, 116
basal metalliferous sediments, 58
drilling techniques, 115
passive margins, 76
Deformation, for oilfields, 35
of sediments, 37
structural, 6
Delta, Niger, 39
Delta toe overthrusts, 39
Diagenetic reaction, 43
Diapirs, salt, 6, 37
shale, 6
Dispute Settlement Procedure, 30
Diver, ambient-pressure, 135, 143, 144, 151
Diving suit, one atmosphere (Jim), 148, 157
Diving systems, one atmosphere, 156
Dredging, 161–168, 176

alluvial tin, 165–168
compensating systems, 163, 176, 177
designs of tin dredges, 167
future trends, 164, 167
methods for sand and gravel, 163–165
on-board processing, 163
suction, 163
tailings disposal, 167
Drilling, controlled offshore, 114
deep water, 6, 113
limit of, 99
slim hole method, 123
uncontrolled offshore, 116

Earthquakes, in continental margins, 75, 76
East Pacific Rise, ferromanganese deposits, 56–64
Economics, development of oilfield, 12, 13
North Sea, 34
oil production, 12–17, 34, 35
Energy policies, national, 3
Engineering, offshore subsea, 99–111
Enhancement, oil yield, 18
research, 19
Environment, effect on manganese nodule mining, 168–170
effect on mineral processing, 161–177
pollution in offshore operations, 179–185
Erosion, by ice, 75, 79
by wave action, 75, 79
Evaporites, 37
in continental margins, 78
Exclusive Economic Zone, 23
Exploitation, continental shelf, 28
resource, 25, 26
see also Minerals and Oil

Fan sands, deep sea, 5, 39
Faulting, of continental margin, 75
Ferrihydrite, *see* Iron oxides
Ferromanganese deposits, accumulation rates, 64
adsorption of metals from seawater, 43
composition, regional variations, 53, 57–62
distribution and composition, 43, 44
distribution in Pacific, 66
epitaxial intergrowths, 51, 65
formation of oceanic, 64–68
geochemistry, 43–73
hydrothermal activity, 67
lanthanide enrichment, 46, 47, 68
mineralogy, 47–52
minor elements, 43–68
Mn/Fe ratios, 44, 45, 53–55
nodules, 43–56
ordered structures, 51
oxyhydroxide component, 62
precipitation and accretion, 43

ridge crest, 56–64
  rôle of microorganisms, 67
  sources of metals, 64
  unconsolidated sediments, 43
  see also Nodules
Fishing, Convention, 22
  rights, 28
Flotation, froth, 172, 173
Flowlines, flexible, 102
  oil, 99, 100
Foraminifera, agglutinating, 67

Galapagos Spreading Centre, 61, 67
Gas, depth of burial, 35
  generation, 3
Gas production, biochemical in shallow
    sands, 37
  deep water, 113–124
  in coral reefs in E Indies, 36
  world, 15
  see also Hydrocarbon
Geology, deep water, 33
Glomar Challenger, 116, 122
Goethite, see Iron oxides
Gravel, offshore, see Sand
Gravity flow, 39, 40, 80
Gravity tectonics, 6
Gulf of Mexico, environmental pollu-
    tion, 179
  oil spills contingency plans, 183
Guyots, 36

High Seas, Convention, 22
  passage, 29
  rights, 28
Horsts, 6
Hydrates, in deep water, 41
Hydrocarbon generation, depth of
    burial, 35
  threshold, 4
Hydrocarbon potential, deep water,
    33–42, 113–124
  of continental slope and rise, 122
Hydrocarbon recovery, see Oilfield
Hydrothermal activity, precipitates, 68
  re ferromanganese deposits, 67

Iceberg, plough marks, 79
Indian Ocean, nodules, 72
Intergovernmental Maritime Consulta-
    tive Organization, 179, 181
International Authority, institutions of,
    27
  powers of, 26
  the Enterprise, 26
International Exploration and Pro-
    duction Forum, 11
  sub-committees, 181, 182
International Law Commission, 24
Intervention systems, subsea, 11
Investment programmes, constraints,
    3
Iron, offshore mining, 161
Iron oxides, amorphous oxyhydroxides,
    63
  Fe-rich nodules, 53
  ferrihydrite, 51
  goethite, 51, 63
  in Red Sea mud, 171
  minerals in nodules, 51

mode of accretion, 64, 65
  see also Ferromanganese deposits

Jigging, on tin dredges, 166, 176

Kerogen, type of, 3

Landlocked states, group of, 29
Lanthanides, enrichment in ferro-
    manganese deposits, 46, 49, 55,
    60, 68
Law of the sea and seabed, 21–31, 161
  Tribunal, 30
  U.N. Conference, 21, 125
Lead, in ferromanganese deposits, 43,
    46, 49, 53, 55, 58
  isotopic compositions, 62
Licence terms, see Oilfield

Madagascar, oil sands, 40
Manganese nodules, 25
  deeper water problems, 184
  effects of marine environment on
    mining, 168–170
  environmental pollution, 179
  material handling, 169
  mining, 125–133
  ship-board processing, 169
  transportation, 169
  see also Ferromanganese deposits
Manganese oxides, birnessite, 48–50, 63
  δ-MnO₂, 48–50, 51, 63, 65
  formation, 65–68
  minerals in nodules, 48–50, 55
  Mn-rich nodules, 53
  todorokite, 48–50, 51, 63, 65, 66,
    73
  X-ray diffraction, 48, 49
  see also Ferromanganese deposits
Manifold, 156
  centre, dry, 103, 106
    underwater, 103, 156
  subsea, 101, 102
Mesoscale features, 92–94
Metal recycling, in ocean, 43
Mid-oceanic ridges, 35, 93
  see also Ridge system
Mineral exploitation, constraints, 161
  deeper water problems, 184
  disposal of waste, 176
  environmental pollution, 179–185
  problems, 161, 162
Mineral processing, 176
  effect of environment, 161–177
  equipment, 176
Mineral separation plant, on tin
    dredges, 166
Minerals, deep seabed, access to, 25
  production of, 25
Mining at sea, problems, 161, 162
Mining, manganese nodules, explora-
    tion and sampling, 125–133
  navigation, 127
Mining system, 128–132
  ocean transport, 132
  the collector, 128, 129
  the lift system, 130, 131
  the mining vessel, 131, 132
Molybdenum, in ferromanganese de-
    posits, 43, 46, 49, 55, 58

Navigation, re deep-sea mining, 127
Navstar Global Positioning System, 127
Nickel-enriched nodules, 53–55, 67
Nickel, in ferromanganese deposits, 43,
    46, 49–51, 58, 126
  sources of, 25, 26
Nine Nations Convention on Civil
    Liability, 182
Nodules, ferromanganese, diagenetic
    precipitation, 65
  from Indian Ocean, 72
  growth rates, 52
  hydration rind dating, 52
  microlaminated structure, 52
  mining of, 125–133
  mode of formation, 65
  nucleation, 65
  precipitation, 43, 65
  radiometric measurements, 52
  see also Ferromanganese deposits
North Sea, continental shelf, 28
  contingency plan, 183
  development, 100, 102, 156–159
  economics, 36, 99
  engineering management, 154
  oilfield peak production, 10
  oil spills, 179
  problems, 153

Offshore Pollution Liability Association
    Ltd, 182
Oil, deeper water pollution problems,
    184
Oil companies, national, 3
Oil exploration, deep water, 3, 4
Oil Industry International Exploration
    & Production Forum, 179
Oil pollution, see Pollution
Oil production, deep water, 3, 113–124
  development of infrastructure, 11
  economic limit, 7
  economics, 12–17, 34, 35
  facilities, floating, 100, 107
  facilities on seabed, 17
  hazards of subsea installations, 104
  in underdeveloped areas, 11
  investments, 16
  potential, 15
  problems, 6–12
  submerged control systems, 101
  technology in deep water, 121
  world, 14
Oil sands of Madagascar, 40
Oil slick, 182
Oil spill, 179, 180
Oil storage systems, surface, 107
Oil yield, see Yield
Oilfield, cost of offshore, 154
  development conditions, 13, 153–159
  development costs, 16
  giant, 34
  licence terms, 13, 14
  production-sharing terms, 13
  royalty/tax terms, 13
  service contracts, 13
  structure, 35
  typical North Sea development
    programme, 156–159
Oilfield systems, 7–11, 99–111
  articulated columns, 9

deviated wells, 10
riser systems, 10
satellite wells, 156
Single Anchor Leg, 105–107
stayed tower, 9
tethered buoyant platforms, 9, 156
well clusters, 9

Pacific Ocean, ferromanganese deposits, 44, 66
Passage, innocent, 28
high seas, 29
Phosphorus, in ferromanganese deposits, 45, 55, 58, 68
Pipe laying, development work, 99
Pipeline repairs, 157
Pipelines, accidents, 181
sea bottom, 108
Plate tectonics, 76
Platforms, multi-part, 105
production, 100, 121
depth limit, 104
floating, 99, 156
tethered buoyant, 7
Pollution, deeper water problems, 184
Louisiana experience, 180
mineral, 179–185
oil, 179–185
Pressure chamber, one atmosphere, 156
Processing of minerals at sea, 161
Psilomelane, 63

Rate of return, commercial, 34
Red Sea mud, deeper water problems, 184, 185
environmental pollution, 179
mining, 171–173
Reflexion profiles, 37, 39
Régimes, for the deep seabed, 25, 26, 35
legal, for mineral exploitation, 125
Research, future of, 29
Reservoir rocks, 35, 36
in sedimentary basins, 4, 5
Revised Single Negotiating Text, 26, 27, 30
Ridge system, ferromanganese deposits, 43, 44, 56–64, 67
Rift system, E African, 77
Rifting of continental crust, 75–79
Rigs, deep water, 6
drilling, semi-submersible, 99
offshore, 153
Riser carrier, disconnectable floating, 119, 120
Riser drilling system, 114
Risers, 156
large, 116
multibore production, 99, 105, 107
with buoyancy, 115
Rises, 33
Romanche Trench, 61

Salt diapirs, 6, 78
Sand and gravel, offshore dredging, 161–165

Seabed Committee of the U.N., 23, 24
Seamounts, 93
enrichment of Co, 54
ferromanganese deposits, 50, 55
Sediment cones, off Indus and Ganges, 40
Sediment ridges, 82
Sedimentary basins, oil in, 3
Sedimentary prism, 4, 5, 36, 40
Pacific, 41
reservoir rocks, 4
Sedimentation, contour current, 82
pelagic, 79, 80, 83
Sediments, deep water of Atlantic slopes, 39
deformation, 37
deposition along continental margins, 78
mode of deposition, 40
Palaeozoic onshore, 37
thickness in Altantic, 37, 38
Seismic exploration, 113
Seismicity of continental margin, 75, 76
Separation system, subsea, 17
Shale diapirs, 6
Shear, in continental margins, 75
Ship's motion, monitoring and simulation, 174–176
SLIKTRAK, 182, 183
Slumping, 75, 79, 80
in Rockall Trough, 80
off Grand Banks, 80
Smectite, Fe-rich, 63
Spar, semi, 107, 109
Spar terminals, 99, 107, 111
Spreading, sea floor, 57
Straits, international, 28
Strontium, in ridge crest sediments, 62
Submarine canyons, 39
erosion, 75, 79, 81, 82, 84
Submersibles, 135–152, 157
applications, 158
battery mass, 158
deep water problems, 145
diver lockout, 135
manned, 135–137, 141
engineering, 135–152
operating depth, 137
problems, 145–147
remotely controlled, 137–140, 142, 158
engineering, 135–152
problems, 147, 148
work tasks, 140
Sulphur, offshore mining, 161

Tectonics, gravity, 6
of oceanic margins, 37
Telechirics, for under-sea work, 152
Terminals, E.L.S.B.M., 99
floating offshore, 99
Spar, 99, 107, 111
Territorial Sea, Convention, 21
its limits, 28
Through flow line, 9, 156, 157
Tin, dredging, 165–168

in SE Asia, 165
in St Ives Bay, 165
offshore alluvial, 161
production, 165
Titaniferous sands, offshore mining, 161
Titanium, in ferromanganese deposits, 46, 49, 55
Todorokite, see Manganese oxides
Tower, stayed, 9
Transportation, behaviour of fine-grained bulk cargoes, 176
of damp nodule cargoes, 169, 170
of minerals, 161
Trenches, oceanic, 75, 76
Truman Proclamation, 21
Turbidite sands, oilfields in, 39
Turbidites, 40
Turbidity currents, 75, 80–82

Underwater contractor, see Contractor
United Nations Conference on Law of the Sea, 21
United Nations Environment Programme, 179, 181
United Nations Seabed Committee, 23, 24
Uranium, in ridge-crest sediments, 61, 62

Vehicle design, underwater, 135
Vehicles, see submersibles
Viking Graben, 77
Volcanic features, deep water, 35
Volcanic island arcs, 75, 76
Volcanic mountain ranges, 75, 76

Waste products, disposal at sea, 161, 176
Water injection, 18, 100
Waves, internal, 96
surface, 97
Welding, explosive, 157
hyperbaric techniques, 108, 157
wet, 157
Well clusters, 9
Well completions , 99
Brunei type, 102
underwater, 100, 101, 121, 157
Well maintenance, 106
Wells, deviated, 10, 100
drilling methods, 118
pressure control, 116–118
remote, 100
satellite, 156
semi-submersible, 105
subsea, 105, 107
water injection, 100

Yield, oil, 18
enhancement, 18
research, 19

Zinc, in ferromanganese deposits, 43, 46, 49, 55, 58
in Red Sea mud, 171